Ancient Chinese Inventions

China has given birth to numerous scientific and technological inventions, and for many centuries led the world in such innovations. Indeed, some of the most important inventions in the history of human civilization originated in China—not least the compass, gunpowder, paper, and printing. *Ancient Chinese Inventions* provides an accessible, illustrated introduction to the many inventions to which China can lay claim, from the aforementioned seminal discoveries to mining technology, the production of silk and ceramics, and brilliant advances in astronomy, mathematics, and medicine.

Introductions to Chinese Culture

The thirty volumes in the Introductions to Chinese Culture series provide accessible overviews of particular aspects of Chinese culture written by a noted expert in the field concerned. The topics covered range from architecture to archaeology, from mythology and music to martial arts. Each volume is lavishly illustrated in full color and will appeal to students requiring an introductory survey of the subject, as well as to more general readers.

Deng Yinke

ANCIENT CHINESE INVENTIONS

CAMBRIDGE
UNIVERSITY PRESS

CAMBRIDGE UNIVERSITY PRESS
Cambridge, New York, Melbourne, Madrid, Cape Town, Singapore,
São Paulo, Delhi, Dubai, Tokyo, Mexico City

Cambridge University Press
The Edinburgh Building, Cambridge CB2 8RU, UK

Published in the United States of America by Cambridge University Press,
New York

www.cambridge.org
Information on this title: www.cambridge.org/9780521186926

Originally published by China Intercontinental Press as
Ancient Chinese Inventions (9787508516899) in 2010

© China Intercontinental Press 2010

This updated edition is published by Cambridge University Press
with the permission of China Intercontinental Press under
the China Book International programme 👁.

For more information on the China Book International programme, please visit
http://www.cbi.gov.cn/wisework/content/10005.html

Cambridge University Press retains copyright in its own contributions
to this updated edition

© Cambridge University Press 2011

First published 2011

A catalogue record for this publication is available from the British Library

ISBN 978-0-521-18692-6 Paperback

Contents

Foreword

China has attracted the attention of the world with its high-speed economic growth. As the country expands its global influence, people have become more interested in this ancient nation. They are eager to learn about its history, which might perhaps provide a clue to the driving force behind its current rapid economic growth. The Western world used to know little about China's past, except perhaps for the four major inventions: compass, gunpowder, paper, and printing. In fact, China led the world throughout much of the history of human civilization with its

The *hun yi* (armillary sphere), one of the major astronomical instruments in ancient China. This sketch of the *hun yi* is by Su Song of the Song Dynasty.

ancient science and technology, and until the middle of the nineteenth century its economy was the largest in the world. The development of science and technology in ancient China was based on the observation and study of the human body and of the world, including the heavens and the earth, leading to the concept of "integration of nature and man." These achievements nourished Chinese culture and civilization, and contributed greatly to human society. The inventions and discoveries of ancient China are rich and numerous.

From the earliest history of human civilization, China has been at the forefront of science and technology, a position achieved in relative isolation. In the Neolithic Period, animal husbandry, farming, construction, pottery, weaving, brewing, and medicine had all developed, albeit at an embryonic stage. The bronze culture of the Shang (1600–1046 BC) and Zhou (1046–771 BC) Dynasties opened a new chapter of civilization, paving the way for the development of the economy, culture, and science and technology. The

Spring and Autumn (770–476 BC) and Warring States (475–221 BC) Periods were a significant time, full of creative enthusiasm, when ancient philosophers competed in exploration of truth. The most important discoveries and inventions in these periods were the production of iron and steel, basic to so many inventions and innovations, which led to a change from bronze culture to iron culture. There were rapid developments in farming, water conservation and handicrafts, and the appearance of the disciplines of astronomy, medicine, mathematics, and agronomy.

The achievements of the Spring and Autumn and Warring States Periods laid a solid foundation for the development of science and technology in ancient China. From then on, though with ups and downs, science and technology continued to advance in China. This was true in the Han (206 BC–220 AD) and Tang (618–907) Dynasties, which were strong and unified with vast territories; in the Wei, Jin, and Northern and Southern (220–589) Dynasties, when the country was temporarily divided; in the Song (960–1276) and Ming (1368–1644) Dynasties, which enjoyed relative stability and prosperity; and in the Yuan (1206–1368) and Qing (1644–1911) Dynasties. The Chinese made great advances

Baopingkou, the opening cut in the east bank of the Minjiang River, is an important part of the ancient Dujiangyan irrigation system.

Lu Ban, a legendary carpenter and inventor.

in the natural sciences as well as in art and literature, history, and philosophy, and invented and created many things necessary for use in daily life. In addition to building on earlier scientific achievements, the ancient Chinese made other discoveries and inventions, such as copper, iron, coal, petroleum, porcelain, silk and cotton cloth, wine and liquor, and Chinese medicine. They also made discoveries in the studies of mathematics, physics, chemistry, and astronomy. They manufactured wonderful machines and devices, created accurate calendars, built the Great Wall and the Grand Canal, as well as presenting the musical theory of "equal temperament." Mozi (c. 468–c. 376 BC), a great philosopher, thinker, and

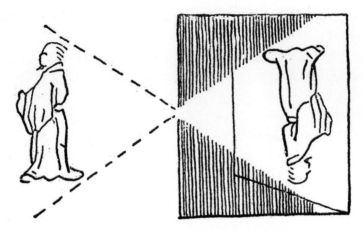

The principle of the camera obscura—light passing through a pinhole in a box forms an inverted and reversed image of the subject on the back of the box.

scientist of the Warring States Period, was the first to discover the principle of the *camera obscura*—light that passes through a pinhole in a box forms an inverted, reversed image of the subject on the back of the box. His discovery was made earlier than that of the Platonic School in Greece. Lu Ban (c. 507–c. 444 BC), a celebrated craftsman, established basic techniques for carpenters without the need for nails or glue. According to legend, Lu Ban once made a seemingly rough and unstable chair and dropped it down from a city wall. When the chair hit the ground, it became a rigid, perfect chair. This story suggests the advanced level of practical techniques in ancient China.

When we discuss inventions and discoveries in ancient China, we must mention the British historian of science Dr Joseph Needham (1900–1995), who firmly believed that China was the cradle of world inventions. In his multi-volume work *Science and Civilization in China*, this British scientist listed the major Chinese inventions and the related time lags between China and the West. These inventions include cast iron, gunpowder, papermaking, compass, movable type printing, porcelain, and the waterwheel. Needham argued that China had scored repeated "world's firsts," and that Francis Bacon (1561–1626) was correct in recognizing that the inventions of gunpowder, paper and printing, and the magnetic compass transformed the antique and medieval worlds into the modern age.

As the American scholar Robert G. Temple argued: "the 'modern world' in which we live is a unique synthesis of Chinese and Western ingredients. Possibly more than half of the basic inventions and discoveries upon which the 'modern world' rests come from China." Even Newton's first law of motion and William Harvey's discovery of blood circulation can be traced to Chinese sources; the Industrial Revolution was made possible in Europe by the preceding European Agricultural Revolution, which was brought about by the importation of Chinese inventions and agricultural techniques and their dissemination.

A foreign scholar, but a recognized authority on the history of science and technology in China, Needham devoted his life to the study of the history of Chinese science, technology, and medicine. Joseph Needham was at first a leading biochemist. He took his Chinese name Li Yuese after the ancient Chinese thinker Li Dan (Laozi) whom he admired greatly. In the mid-1930s, Chinese students at Cambridge awakened Needham's interest in China, and in 1939 he began to learn the Chinese language and to study Chinese history of science. During the War of Resistance against Japan (the second Sino-Japanese War), Needham went to China as Head of the British Scientific Mission and was later Scientific Counselor to the British Embassy at Chongqing, then the "acting-capital" of China. Under the auspices of the British Council, Needham established the Sino-British Science Cooperation Office (SBSCO), and met many Chinese scientists.

While supporting the Chinese resisting Japanese aggression, Needham also began a new phase in his study of the Chinese history of science. Needham had a deep love for the enterprising spirit and great achievements of ancient Chinese scientists, and he devoted all his energy to his study. He read extensively in the Chinese classics, visited the cultural sites of various Chinese dynasties, and even rode on horseback to undertake field studies in northwest China. After the founding of New China in 1949, Needham visited China eight times, traveling widely, collecting large quantities of historical materials on Chinese history of science, and obtaining first-hand knowledge of China's political, economic, cultural, and scientific development. He was

Dr Joseph Needham.

one of the founders of the Britain-China Friendship Association and served as its President in the 1950s. In 1954 he published the first volume of his monumental work *Science and Civilization in China*, with the help, among others, of Dr. Lu Gwei-djen, who awoke his interest in Chinese history. Needham was the only British scientist who had the honor of being both a Fellow of the Royal Society and a Fellow of the British Academy. He was also one of the first foreign academicians of the Chinese Academy of Sciences.

While admiring the brilliant achievements of science and technology in ancient China, Needham could not but wonder why modern China had a rather poor record in science and technology. This led to Needham's "Grand Question" or the "Needham Puzzle"—although it should be noted that Chinese scholars had also raised this question in 1915. The answers to this question are numerous and complex.

For a long stretch of time Chinese bureaucracy had promoted and protected the productivity of science. But in the mid- to late Ming Dynasty, feudal rule was strengthened and the country was closed to the rest of the world. This hindered the development of science and technology severely. In the meantime, Europe was experiencing its Renaissance. For two to three hundred years thereafter, science and technology in China remained in a backward state, especially when the nation became yet more isolated under Emperor Yongzheng (r. 1723–1735) of the Qing Dynasty. Yet it was in this same period that the Industrial Revolution was taking place in Europe. As science and technology in Europe advanced at an ever-increasing pace, their counterparts in China were hindered by rapidly worsening political and economic conditions.

The closed-door policies of the Ming and Qing Dynasties undoubtedly stifled the development of science and technology in China. In the absence of exchange with the outside world there was no understanding of world science or import of ideas from abroad. In the meantime, the feudal, high-handed rule and repression of free thinking prevented enterprising exploration.

After the extraordinary advance of global science and technology in the mid-nineteenth century, China was subjected to aggression and suppression by imperialist powers, which in turn controlled its economy and extorted huge war reparations from the Qing government. The nation was depleted, its scientific and cultural undertakings rapidly declined, and its earlier glory was all but lost.

Science and technology in ancient China also suffered from another deficiency: lack of exchange and distribution of ideas, and an absence of governmental guidance and institutional protection. Scientific research was quite often the endeavor of an individual person, therefore it was not to meet the demands of society, and findings usually disappeared with the death of the scientist. For instance, twelve tone "equal temperament," developed by Zhu Zaiyu (1536–1611) of the Ming Dynasty, is a great achievement in music, but his findings were almost forgotten and there was no successor to carry on his research. Emperor Kangxi (r. 1661–1722) of the Qing Dynasty was a great mathematical enthusiast and made algebra calculations as a pastime, but that was only his personal hobby. The emperor never used his love of mathematics to promote and develop science and technology.

Ancient China had an extreme bias for arts and against science. Ancient Chinese learning included only literature, philosophy, history, and language. Science and technology, which were needed in production and daily life, were regarded as something trivial and vulgar. Intellectuals in ancient China had a very poor understanding of the natural sciences. Among those who had contributed to the development of science and technology in ancient China, there were very few who were funded and rewarded by the government, with the exception of a few imperial historians who made astronomical and calendar-related findings. Students of the whole nation devoted their efforts to the study of classics and stereotyped writings. All the schools were for the arts and there were hardly any institutions for the training of scientific personnel. Although the country had a large population and a large number of students, very few of them devoted themselves to natural sciences and technology. With the implementation of imperial examinations, by which government officials were chosen according to the quality of the essays they had written, the shortage of scientific personnel became still more acute.

In addition, the few Chinese who did study science quite often focused their efforts on questions related to philosophy and practical techniques rather than on issues of basic sciences such as physics, chemistry, and biology. They also had little interest and experience in purposeful scientific experiments. As European scientists experienced an explosive advance in science and technology in the past two or three centuries, their Chinese counterparts were left far behind.

A review of inventions and discoveries in ancient China should help people recognize the wisdom and creativity of the Chinese, as well as the internal dynamic behind China's rapid economic, cultural, scientific, and technological growth over the last thirty years. The years of reform and opening up to the rest of the world are but a second in Chinese history, and China's economy has expanded greatly. The gap between China and advanced Western countries has been reduced, and in some fields the country has moved to the forefront.

Deng Xiaoping, the chief architect of China's reform and opening up, said, "Science and technology constitute a primary productive force." The Chinese government has recognized the vital role of science and technology in promoting production, and regarded it as a basic state policy to utilize science for the development of the country. China has made substantial progress in long-term planning for the development of science and technology, building infrastructure facilities, training scientific personnel, and exchanging information. The nation is losing no time and is forging ahead to catch up with more advanced countries, and the future seems bright for the renewal of China, especially in the fields of science and technology.

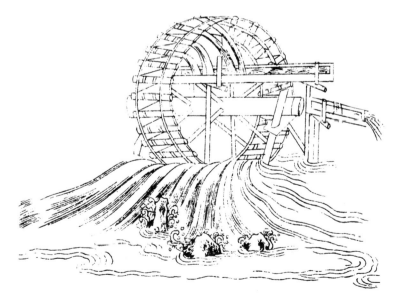

The Four Major Inventions

For a thousand years the Silk Road was a major route of exchange between China and the rest of the world.

Are the Four Major Inventions the Most Important?

The designation of paper, printing, gunpowder, and the compass as the four major inventions was not made by the Chinese, but by foreign scholars with an interest in Chinese civilization.

Derk Bodde (1909–2003), an American scholar, argued that, "Without paper and printing, for example, we should still be living in the Middle Ages. Without gunpowder, the world might have been spared much suffering, but on the other hand the armored knights of medieval Europe might still reign supreme in their moated castles, and our society might still be held in feudal servitude. Nor would the building of the Panama Canal or of Boulder Dam have been possible! And finally, without the compass, the great age of discovery might never have come, with its quickening of European material and intellectual life, and its bringing to knowledge of worlds hitherto unknown, including our own country."

Chinese inventions have made a major contribution to world civilization.

Similarly, the early modern English philosopher Francis Bacon wrote, "Again, it is well to observe the force and virtue and consequences of discoveries, and these are to be seen nowhere more conspicuously than in those three which were unknown to the ancients, and of which the origin, though recent, is obscure and inglorious; namely, printing, gunpowder, and the compass. For these three have changed the whole face and state of things throughout the world; the first in literature, the second in warfare, the third in navigation; whence have followed innumerable changes, insomuch that no empire, no sect, no star seems to have exerted greater power and influence in human affairs than these mechanical discoveries."

The dawn of the Renaissance in Europe coincided with the availability of paper for printing.

However, it is doubtful that consideration of these four great inventions can reflect the achievements of scientific and technological inventions in ancient China precisely. The four inventions were regarded as the most important Chinese achievements in science and technology mainly because they had a prominent position in the exchanges between the East and the West, and acted as a powerful dynamic in promoting the development of capitalism in Europe. In fact, the ancient Chinese achieved much more than the four major inventions: there were major developments in farming, iron and copper metallurgy, exploitation of coal and petroleum, machinery, medicine, astronomy, mathematics, porcelain, silk, and wine making. The numerous inventions and discoveries related to people's livelihoods and daily life advanced China's society. Many are at least as important as the four major inventions, and some are arguably even greater. Before touching on others, however, let us begin with the famous four.

Compass

The compass is called *zhi nan zhen* in Chinese, literally meaning a needle pointing south (south is the primary direction in China, just as north is in the West). As a device used to determine geographic direction, the compass usually consists of a magnetic needle mounted or suspended and free to pivot until aligned with the magnetic field of Earth, and a dial. The Chinese classics mention another device for indicating direction, *zhi nan che* (vehicle pointing south), and legends claim it was invented by Huangdi (the Yellow Emperor), the earliest ruler of the Chinese nation. A more reliable record states that scientist Ma Jun of the Three Kingdoms Period (220–265) restored this earlier invention, but it gives few details of the device. Researchers believe that the vehicle was equipped with complicated gears and clutches, and it is possible that the hand of the wooden figure on the vehicle pointed south wherever the vehicle moved.

The earliest primitive magnetic compass in China, *si nan*, was probably invented during the Warring States Period, for the use of *si nan* was recorded in several Chinese classics of that period. The device consisted of a spoon cut out of lodestone and a bronze plate, the surface of both being smooth enough for the spoon to turn easily on the plate, on which twenty-four directions were marked. To use the device, the spoon was put in the center of the plate and turned slightly. The spoon would move around and finally the handle would point south, because of magnetism. But lodestone was not easy to obtain and the spoon and plate were too heavy to carry around. Later inventors artificially magnetized iron needles or iron pieces in other shapes, and developed the familiar magnetic compass.

Scientist Shen Kuo (1031–1095) of the Northern Song Dynasty wrote in his *Dream Pool Essays* that geomancers pursued their art by rubbing a lodestone against a steel needle, thus causing

A model of the *zhinan che*, by scientist
Ma Jun of the Three Kingdoms Period.
The wooden figure, operated by gears
and clutches, points south, whichever
way the vehicle moves.

Model of a floating compass. The magnetic
"fish" comes to rest pointing north/south.

Model of a suspended compass. A magnetic
needle is suspended above a compass dial
by a single thread of silk.

This illustration shows the Ming Dynasty eunuch Zheng He on his voyage to the west, a journey facilitated by the use of the magnetic compass.

the needle to point south. Such a needle, he added, can then be floated on water, put on the edge of a nail or bowl, or, best of all, suspended from a thread. He noted further that the needle never points exactly to true south, but always deviates slightly. The knowledge shown here of the principle of magnetic deviation almost certainly proves that the compass had been known and studied by the Chinese before Shen Kuo's time.

It is thought that China was the first country in the world to use the compass in navigation, and its spread to the West had a tremendous impact on world civilization. In the early twelfth century the Northern Song government sent a large fleet to Korea. A book that recorded this voyage says that the fleet "observed the Dipper to determine the direction, and when it was overcast the compass was used." The use of the compass ended the dependence on astronomical observation in navigation. In Europe the compass was first mentioned in a French poem of 1190, but its application to navigation was mentioned only later.

Gunpowder

Gunpowder, a preeminent representative of science and technology in ancient China, was invented by alchemists. Its ingredients include sulfur, saltpeter (potassium nitrate), and charcoal. The invention was made in the Tang Dynasty, and improved during the Song Dynasty. The Chinese for gunpowder, *huo yao*, literally means fire drug, or the drug that fires. In *Shennong's Herbal Classic*, believed to have been written in the Han Dynasty, sulfur and saltpeter were listed as important drugs. The most famous figure of Chinese *materia medica*, Li Shizhen (1518–1593) of the Ming Dynasty, writes in his *Bencao Gangmu* (*Compendium of Materia Medica*) that gunpowder could be used to treat infections of the skin, and to kill pests. The Chinese alchemists failed to produce an elixir of life, but they did accumulate rich knowledge of chemistry. Since they knew that sulfur was quite chemically active, toxic, and inflammable, they mixed it with saltpeter and charcoal before heating. Saltpeter, as a strong oxidant, would cause partial combustion of the sulfur, thus reducing its toxicity

Alchemists experiencing an explosion. Although they failed to find the elixir of life, alchemists accumulated a rich knowledge of chemistry.

Gunpowder was widely used to make firecrackers for celebrations.

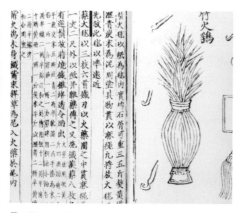

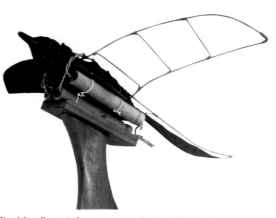

The Northern Song Dynasty *Wujing Zongyao* (Compendium of Military Classics) included formulae for gunpowder.

The Ming Dynasty fire crow was a firebomb filled with gunpowder, which could fly several hundred meters and explode when it landed in an enemy camp.

and inflammability. In his *Classic of Alchemy*, medical scientist Sun Simiao (581–682) of the Tang Dynasty called this method "controlling sulfur."

While trying to "control sulfur," Sun found that a mixture of certain proportions of sulfur, saltpeter, and charcoal would result in an explosion. When this knowledge was passed to the defense-related industry, craftsmen made further studies and tests to find out the best ratios of the mixtures, thus turning the material into a practical explosive that burned and exploded in a sealed container.

During the Northern Song Dynasty (960–1126), Zeng Gongliang, an engineering expert, recorded three formulas of gunpowder in his *Wujing Zongyao* (*Compendium of Military Classics*, 1044): one for cannons, another for poison smoke bombs, and the other for fireballs, with detailed ratios of ingredients. The formulas show that the proportion of saltpeter was larger than the combination of sulfur and charcoal, which is similar to the ratios of modern black powder. In the Northern Song Dynasty various weapons that used gunpowder were developed, most notably the "fire gun," said to be the precursor of modern firearms. In the thirteenth century the Yuan Empire sent troops on a military expedition to Central Asia, and gunpowder and firearms thus reached the Arab world and then Europe.

Paper

Paper has been a major medium of recording, transmitting, and storing information for a long period of human civilization, especially for the works of philosophers, writers, statesmen, scientists, and historians. Books, newspapers, letters, accounts, and archives, for example, all depend on paper.

In China, ancient characters were first cut on tortoise shells and animal bones, and cast on bronze wares. Later, in the Spring and Autumn and Warring States Periods, bamboo, wood, and silk were used for writing.

But bamboo or silk both had their defects. Bamboo strips were clumsy to handle, and a book would require a large number. For instance, Hui Shi (390–317 BC), a noted scholar of the Warring States Period, used to take with him five carts loaded with books on bamboo strips. It took two people to carry a memorandum on bamboo strips submitted by ministers to Emperor Wudi (r. 140–87 BC) of the Han Dynasty, who took two full months to read it. The other material, silk, was too expensive for ordinary people to use for writing. A lighter and cheaper material was needed to replace them.

The ancient Chinese wrote inscriptions on animal bones and tortoise shells. This piece dates from the Shang Dynasty.

In fact, paper emerged in the Western Han Dynasty (206 BC–25 AD): a piece of material unearthed at Baqiao in the suburbs of Xi'an

Silk and bamboo strips were both used for writing. The former was expensive and the latter rather clumsy, so alternative materials were sought.

Portrait of Cai Lun.

in 1957 was proved to be paper made of hemp and ramie fibers; and in 1986 a map unearthed in Tianshui of Gansu Province was also proved to be drawn on a piece of paper made of fibers from silk and hemp. But the earliest paper was rough and not quite suitable for writing, and the materials were not easy to obtain.

Cai Lun (?–121), a eunuch of the Eastern Han Dynasty (25–220), made a revolutionary innovation in papermaking. He used bark, hemp fiber, broken fishnets, and rags as raw materials. The materials were soaked, cut into pieces, boiled with plant ash, washed, and ground with a pestle in a mortar. The mixture was then poured evenly onto a flat surface to dry, or taken off and baked to become paper. The process was later improved, and a mold of bamboo strips was used to dredge paper from a suspension. The materials were also expanded to include bamboo, reed, rattan, straw, and various kinds of bark. With better materials and improved processes, some very fine varieties of paper were made in later dynasties.

The process of making paper.

In quite a number of places in southern China, such as Dengcun in the suburbs of Sihui City in Guangdong Province, a place with rich supplies of bamboo, local people still make paper from bamboo using Cai Lun's precise process, and this high-quality bamboo paper sells well in Southeast Asian countries. A visit to Dengcun gives visitors a journey into history to see how Cai Lun made paper.

The Chinese papermaking technique was first introduced to Korea and Vietnam, then to Japan in the seventh century, to Arab countries in the eighth century, to Europe in the mid-twelfth century, and four hundred years later to South America. In the two thousand years from the second century BC to the eighteenth century, China led the world in papermaking, making immeasurable contributions to the dissemination and recording of knowledge and the accumulation and exchange of culture.

Printing

The trinity of paper, printing, and books has made a huge contribution to the growth of human civilization, dissemination

of scientific and cultural knowledge, and promotion of friendship between peoples. Printing has a long history. The techniques discussed here include both block printing and movable type printing. Block printing was probably invented between the Sui (581–618) and Tang Dynasties, based on the technique of transferring text and pictures cut in relief on seals and stone pillars to other surfaces that was developed in the Spring and Autumn and Warring States Periods. The invention of paper and the improvement of ink led to the advance of block printing.

Bi Sheng, who invented moveable type printing.

In block printing, the text was first written in ink on a sheet of fine paper; then the written side of the sheet was applied to the smooth surface of a block of wood, coated with a paste that retained the ink of the text; thirdly, an engraver cut away the blank areas so that the text stood out in relief and in reverse. To make a print, the wood block was inked with a paintbrush, a sheet of paper spread on it, and the back of the sheet rubbed with a brush.

The earliest existing work produced by block printing in China is the *Jin Gang Jing* (*Diamond Sutra*) printed in 868 during the Tang Dynasty. In the Five Dynasties (907–960), government-run cultural institutions engraved and printed ancient classics on a large scale, and non-governmental publishing was also quite popular. In the Song Dynasty

Chi Sheng Guang Jiu Yao Tu (Tejaprabha Buddha and the Nine Planets), a block-printed picture from the Liao Dynasty (907-1126). This was found in 1974 in an old tower in Yingxian County, Shanxi Province.

the Buddhist work *Da Zang Jing* (*Tripitaka*) was printed and a total of 130,000 wood blocks were engraved for the project. The technique of block printing first spread to Japan and Korea, then to Egypt in the twelfth century, and to Europe in the fourteenth century. In Japan the *Daranikyo Sutra* printed in 770 is the oldest extant work from block printing. As block printing was so complicated and difficult, it would take several years to print a book, and the engraved blocks required a large storage space.

Movable type printing was a later invention. Bi Sheng (?–c. 1051), a worker in a printing shop in the Song Dynasty, devoted great efforts to his invention of movable type printing. The principle of Bi's invention is the same as that of typeset printing widely used in the twentieth century. In his *Mengxi Bitan* (*Dream Pool Essays*), Shen Kuo wrote about Bi's movable type printing invented in around 1041–1048. Bi composed texts by placing the types side by side on an iron plate coated with a mixture of resin, wax, and paper ash. Gently heating this plate and pressing the types with a smooth plate to ensure they are on the same level, and then letting the plate cool, meant the type was fixed on the plate with melted wax and became a piece of print block. Once the impression had been made, the type could be detached by reheating the plate. Bi prepared two iron plates to be used in turn to speed up the whole printing process. He also prepared different numbers of types for characters according to their frequency of use in texts, and arranged them in an orderly way to

Imitation of the clay movable type of the kind invented by Bi Sheng.

facilitate composition. Shen noted that this technique was most efficient in printing several hundred or several thousand copies.

After Bi Sheng, other people invented types cut out of wood. In around 1313 Wang Zhen, an agronomist of the Yuan Dynasty, printed his work *Nong Shu* (*Treatise on Agriculture*) with movable wood types, and wrote about his innovation in an appendix to the treatise. He also invented horizontal compartmented cases that revolved about a vertical axis to permit easier handling of the type. Wang tested his technique, and printed in a month one hundred copies of the 60,000-character *Jingde Xian Zhi* (*Jingde County Annals*), which was quite a remarkable achievement at that time. Movable wood type printing was quite popular in China after Wang Zhen, especially in the Ming and Qing Dynasties. In 1773 the Qing government under Emperor Qianlong printed *Wuyingdian Juzhenban Congshu* (*Wuying Palace Series of Treasured Books*), with 138 titles and more than 2,300 chapters. It was the largest project of movable wood type printing in Chinese history, and took more than 250,000 types in different sizes cut out of jujube wood. After wood types, other researchers invented types made of metal: copper type in 1488 in the Ming Dynasty, and lead type in the early sixteenth century.

The technique of movable type printing was introduced to Japan and Korea in the fourteenth century. In the West, German printer Johannes Gutenberg is credited with the invention of typographic printing in the mid-fifteenth century.

A model of horizontal cases for moveable type printing, a device invented by Wang Zhen. The cases, containing compartments, revolved about a vertical axis.

Other Major Inventions (I)

Bronze chariot and horses unearthed at the mausoleum of Emperor
Shihuang of the Qin Dynasty.

As the four major inventions enjoy such prestige, specialists in the history of science and technology in China inevitably emphasize their own areas of study and seek to name a fifth major invention. Thanks to Joseph Needham's strenuous work, new historical and archaeological discoveries, and the efforts of Chinese historians of science, inventions and discoveries in ancient China are attracting growing attention worldwide. It is quite understandable that historians of science have sought to name a fifth major invention, but this is a mission impossible. The ancient Chinese had a number of significant achievements in metallurgy, extraction of petroleum and coal, farming, manufacture of cotton and silk fabrics, traditional Chinese medicine, astronomy and calendar, mathematics, pottery and porcelain, musical tuning, architecture, mechanical engineering, and in water conservation. Several of these achievements may have scientific and practical values that can compare with those of the four major inventions. Here we take a look at some of them.

Bronze axe head with embedded iron edge, unearthed from a Shang Dynasty tomb in Gaocheng, Hebei Province.

Iron and Steel Smelting

The use of iron was an important driving force in human history. Archaeological findings in the twentieth century proved that in China there was a boom of ironware making in the late Spring and Autumn Period, and by the late Warring States Period the industry reached a peak. Successive finds of iron artifacts indicate that ancient China had developed techniques for iron production that were more advanced than its foreign counterparts.

The world's earliest pig iron was found in central Asia. But as the furnaces used were quite small with a weak blast, the output was only low-quality iron. Ancient China began to lead the world in this field in the late Spring and Autumn Period, and the main technique for iron smelting in the Warring States Period involved vertical furnaces. In the Han Dynasty, with the craftsmanship greatly improved in state-run workshops, larger and larger furnaces were used.

The main part of the bellows was a leather bag named *tuo*, and several such bags were linked together to make a powerful bellows with a greater output of air, raising the temperature in the furnace up to 1,200 degrees. At first, manpower or animals were used to drive the bellows. Historical records report that it took a hundred horses to drive the bellows and up to a thousand workers to fill the furnace with ore. This must have been a truly

The Iron Lion of Cangzhou, Hebei Province, cast in the Five Dynasties.

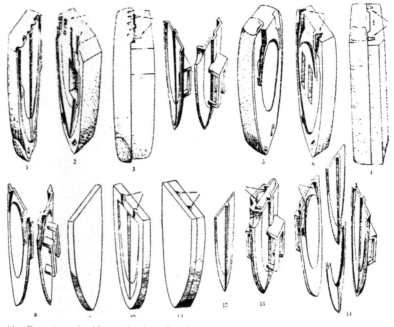

Han Dynasty method for casting iron plowshares.

magnificent scene. As the industry developed, manpower or animal power was no longer strong enough for the blower, and the hydropower bellows was invented. The scientist Wang Zhen of the Yuan Dynasty described and illustrated the structure of the hydropower bellows in his *Treatise of Agriculture*. The machine was installed close to a rapid stream, with a transmission mechanism linking a water wheel with the bellows. As the water flow drove the wheel, the bellows sent a strong blast into the furnace. The driving, transmission, and operating mechanisms of the hydropower bellows were a perfect match for the unique furnace. In Europe, the hydropower bellows was first introduced in the twelfth century.

Malleable cast iron was another major invention of ancient Chinese ironworkers. Simple cast iron is not suitable for forging fine iron implements, because of its brittleness. Malleable cast iron can be divided into white-heart malleable iron and black-heart malleable iron, which are the results of different heat treatment methods. White-heart malleable iron has stronger rigidity and strength, while black-heart malleable iron has greater shock-resistance. Western works of the history of metallurgy all attribute the two

Iron hoe of the Warring States Period. Iron tools boosted agricultural production at that time.

kinds of malleable iron to European and American inventors: the white-heart one was said have been invented by the French in 1722, and is often referred to as "European malleable cast iron;" black-heart malleable iron by the Americans in 1826, hence the name "American malleable cast iron." But Chinese historical records and evidence provided by archaeological findings show that ironworkers in ancient China had in fact invented the techniques for making these two kinds nearly two thousand years earlier. The key to the techniques is to heat ordinary cast iron at high temperatures for a long time, so that the content of carbon changes. When the carbon content is between those of cast iron and steel, the iron becomes malleable and retains certain rigidity. A spade unearthed in Luoyang (Henan Province), and a hexagonal hoe excavated from the ancient copper mine in Daye (Hubei Province) were both made of white-heart malleable iron.

Another major invention of the ancient Chinese iron and steel industry is the "stirring method" for making steel. This was quite simple and effective. Malleable cast iron with high carbon content was heated to a molten state, and iron ore powder was added and stirred continuously. The carbon in the cast iron was exuded and oxidized, resulting in medium or low carbon steel. When the process was continued, removing more carbon, the product would be wrought iron, which had much lower carbon content.

This technique started in the Western Han Dynasty and was recorded in the *Taiping Classics* of the Eastern Han Dynasty. Coins made of "stirred" steel and a furnace for "stirring" steel have been unearthed at the site of an ancient iron workshop in Gongxian County, Henan Province.

With the development of iron and steel smelting, techniques for making weapons also made remarkable progress. Chinese historical records and legends refer to the names of great ironworkers. Among these are Ganjiang and Moye, a couple living in the Spring and Autumn Period. The King of Chu ordered that the husband and wife should make two double-edged swords, named Ganjiang and Moye, within three years. The couple completed the task, and the swords were so sharp and wonderful that they could even cut hairs when the hairs were blown toward them, and cut iron as easily as cutting mud. Knowing that the tyrant king would kill them because of some delay in making the swords, Ganjiang kept the Ganjiang sword, leaving it to his son and hoping that the son would revenge his parents. In the end the son did kill the tyrant king.

Copper Smelting and Bronze Ware

The earliest copper objects were found in Turkey, dating back 9,000 years. The earliest bronze ware in China was made more than 6,000 years ago, in the period of the Yangshao Culture. The bronze artifacts unearthed from the ruins of the Longshan Culture, dating back to 2500 BC, were made of an alloy of copper and tin, in some cases with a certain amount of lead. Some fragments of these articles proved them to be of even thickness. These wares were cast by split molds, which show that the bronze industry was quite well developed. Later in the ruins of the Qijia Culture, archaeologists unearthed articles made of bronze, brass, copper, bronze mirrors, crucibles, and residues from copper smelting. The findings proved that in 2000 BC, bronze was widely made and a special kind of bronze had been developed for making mirrors. During the Xia Dynasty (2070–1600 BC), bronze-making technology developed in a comprehensive way. A three-leg

Bronze *jue* (three-legged wine vessel), unearthed at the Eritou site in Yanshi, Henan Province.

wine vessel unearthed from the Erlitou site in Yanshi of Henan Province is a representative of Xia bronze ware. The vessel is made of a copper-tin alloy, with 92% copper and 7% tin, and cast in complex molds. Also unearthed were various molds, which show that bronze smelting and casting were quite common at that time. The Xia Dynasty belonged to the Bronze Age.

The heyday of the Bronze Age in China lasted more than 1,600 years, from the Shang Dynasty, Western Zhou Dynasty (1046–771 BC), and the Spring and Autumn Period, to the early years of the Warring States Period. The bronze ware of the time included ritual and musical instruments, weapons, and miscellaneous articles. The musical instruments were mainly used in sacrificial activities in ancestral temples. The ritual instruments were used in various rites—some were exhibited at temples, some were used for dining and washing, and some were to be buried with the dead as funeral objects. Ritual bronze ware was of a divine nature and not used in daily life. Most of the bronze items were ritual articles, featuring excellent craftsmanship. The ritual and musical instruments were the best of bronze ware in ancient China. The ritual items included those for cooking and dining, wine and water containers, and also divine figures. The weapons were mainly *ge* (dagger-axe with a long handle), *shu* (long-poled edged weapon), *ji* (halberd), and *mao* (spear). The bronze ware was decorated with various mysterious patterns, often inspired by images of animals, such as dragon, tiger, ox, goat, deer, phoenix, and birds, as well as human beings. In the Western Zhou Dynasty, the patterns on bronze ware became less mysterious, but dragon and phoenix remained the main themes used, though in varied forms.

Ancient Chinese bronze ware was characterized by fine craftsmanship and tended to be large in scale. Bronze masters developed complex pottery molds, made of selected clays, and cut with various patterns. The wares were either cast as a whole piece or in parts which were then joined together in further

Bronze ware with dragon and phoenix design.

Simuwu Fangding (rectangular caldron), unearthed at the site of the Yin ruins in Anyang, Henan Province.

casting. Later the dewaxing process was developed, which made the casting of separate parts unnecessary. This process was indeed a breakthrough in bronze metallurgy. The ancient Chinese also developed the technique of embedding other materials in bronzeware for aesthetic purposes. The materials used included calaite, jade, siderite, and copper, and during the Spring and Autumn and Warring States Periods, gold and silver were also used. The bronzeware of the Western Zhou Dynasty is noteworthy for the inscriptions used. The *Dayu Ding* caldron of King Kangwang had an inscription of 291 Chinese characters, and the *Maogong Ding* caldron, an inscription of 497 characters. These inscriptions not only testify to the fine craftsmanship of the bronze-ware makers, but they are also important historical records in themselves, complementing relevant historical documents.

Quite a number of substantial and heavy bronze artifacts of the Shang and Zhou Dynasties have been unearthed. The most well-known among these is the *Simuwu Fangding* (rectangular caldron), unearthed at the Yinxu site in Anyang (Henan Province). A vessel from the late Shang, it is 133 cm high, 111 cm long and 79 cm wide, with a weight of 832.84 kg. Its design is solemn and elegant, with patterns on all sides of its exterior. It is the heaviest ancient bronze article in China and the largest in the world, and the image of this caldron is often used as a symbol of ancient Chinese civilization.

In the Eastern Zhou (770–256 BC) period, the bronze smelting technique developed rapidly. *Kao Gong Ji*, a special work on craftsmanship in the Qi State, recorded in detail the proportions of six bronze alloys for making bells and caldrons, axes, halberds, swords, arrows, and mirrors. As wars were frequent, the making of weapons developed rapidly. Especially well-known are swords of the Wu and Yue Kingdoms, and masters of sword making

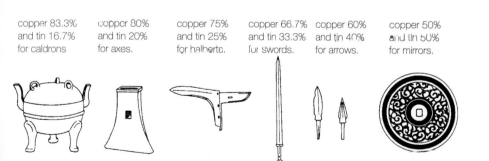

| copper 83.3% and tin 16.7% for caldrons | copper 80% and tin 20% for axes. | copper 75% and tin 25% for halberts. | copper 66.7% and tin 33.3% for swords. | copper 60% and tin 40% for arrows. | copper 50% and tin 50% for mirrors. |

Six formulae for alloys of copper and tin used for caldrons, weapons and mirrors, as recorded before the Qin Dynasty in *Kao Gong Ji*, an important work of science and technology.

such as Ganjiang and Ou Yezi. Some swords that have been buried underground for more than 2,000 years are still quite sharp and can easily cut folds of paper. The sword of Gou Jian, King of the Yue, was unearthed in 1965. The surface of the sword was covered with anti-rusting diamond-shape patterns, the result of certain chemical treatments, and when unearthed was as rust-free and shiny as new.

Gou Jian's sword, unearthed in 1965.

Another treasure of ancient Chinese bronze ware that has amazed the world is the two sets of chariots and horses unearthed at the mausoleum of Emperor Shihuang of the Qin Dynasty in Lintong, Shaanxi Province. The No.1 or lead chariot was drawn by four horses, with the driver seated under a large umbrella. The No.2 chariot, also drawn by four horses, measures 3.17 m long and 1.06 m high. The two sets of chariots and horses are the most complicated bronze works ever to have been unearthed. The half-size chariots and realistic horses were constructed using delicately cast bronze parts, and the chariots and horses were colorfully painted, and decorated with many gold and silver ornaments.

Copper smelting in ancient China was highly developed, the smelters turning out not only bronze, but also copper, brass, and copper-nickel alloys; a large-scale production unrivalled in the world. The site of an ancient copper mine at Tonglu Mountain in Daye, Hubei Province ran 2 km from north to south and 1 km from east to west, covering an area of 140,000 sq m. The mine was more than 50 m underground, with shafts, tunnels and working faces, forming a complete mining system. Primary dressing of ores was carried out in the tunnels.

Tunnels in the site of an ancient copper mine, Tonglu Mountain.

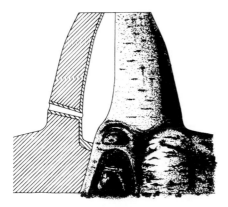

Diagram of restored copper-smelting furnace, Tonglu Mountain, Spring and Autumn Period.

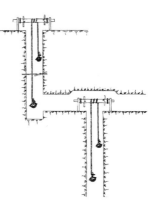

Method used for lifting copper ore in the ancient mine in the Tonglu Mountain.

Extraction and Application of Petroleum

Petroleum is now the lifeblood of industry and transportation, as well as the main source of power in daily life. Ancestors of the Chinese nation were the first in the world to discover, extract, and use oil. *Yi Jing* (*Book of Changes*), which is one of the earliest Chinese classics, dating back more than 3,000 years, states: "There is fire in the marshes." In *Han Shu* (*History of the Han Dynasty*), it was recorded: "There is inflammable water on the Weishui River in Gaonu County." Gaonu County was situated in today's Yan'an, and the Weishui is a branch of the Yanshui River, and it was in this area that petroleum was first discovered, extracted, and used in ancient China. Fan Ye, a historian of the Jin Dynasty (265–420), writes in his *Hou Han Shu* (*History of the Later Han Dynasty*): "To the south of Yanshou County seat there is a mountain, where a spring in the rocks gushes black water that is not potable, and local people call it stone paint." In the Tang Dynasty, writer Duan Chengshi gives a more detailed description of petroleum in his work *Youyang Zazu* (*Miscellanies of Youyang*): "In the Weishui River of Gaonu County there is a greasy matter like paint flowing on the water. Local people get it to grease their wagons and burn it for lighting."

In the eleventh century, scientist Shen Kuo wrote much more about petroleum in his *Mengxi Bitan* (*Dream Pool Essays*), describing its characteristics, applications, and prospects. Shen had inspected the petroleum output in Fuyan County when he served in the government of Yanzhou. He writes: "In Fuyan County there is *shi you*. I know its soot can be used to make ink stick. I tried and made an ink stick, which is black and shining, far better than that made of pine wood soot... This thing will surely have wide applications in the world. As the rock oil abounds in the earth, its supply is ample, unlike pine wood, which may someday be exhausted." He named

Digging the opening of a well.

Laying stone rings.

Digging the well.

the ink "Yanzhou Stone Liquid." Shen Kuo was the first scientist to christen petroleum as *shi you* (literally meaning rock oil, the term for petroleum still used in the Chinese language), and he accurately predicted its future of many applications.

China was also the first to refine oil. In the Northern Song Dynasty (960–1126) a workshop was set up in Kaifeng to produce refined oil for the military. The troops filled iron cans with the refined oil and threw them toward the enemy troops, causing a fire—effectively the world's first "fire bomb." China was also the first to drill oil wells. The ancient Chinese first found natural gas when they excavated rock salt. Zhang Hua of the Jin Dynasty wrote in his *Bowu Zhi* (*Records of Curiosities*) how people in Zigong, Sichuan excavated natural gas and used it to boil a rock salt solution. In 1041, a well with a diameter about the size of a bowl and several dozen feet deep was drilled for salt production. The tool used to dig the well was a "circular blade," the principle of which is similar to that of modern drilling machines. The thirteenth century saw the first well drilled for oil production; in the late Ming Dynasty a 100 m well, the first in the world, was drilled in Leshan, Sichuan.

Song Yingxing (1587–1666), a Ming Dynasty scholar, wrote in detail the methods of extracting petroleum in his *Tiangong Kaiwu* (*The Exploitation of the Works of Nature*). Song's book arrived in Japan in the sixteenth century and in Europe in the eighteenth century. Outside China, the Russians drilled their first oil shaft in 1848, and the Americans in 1859.

Discovery and Mining of Coal

In ancient China coal was called stone charcoal, black wood, black gold, and flammable stone. The earliest record of coal is in the ancient geographical work *Shan Hai Jing* (*Classic of Mountains and Seas*). The ancient Chinese began using coal as an energy source quite early, and by the Han Dynasty there was a coal mining industry. Coal ashes and briquettes were unearthed in the ruins of residential housing in Fushun in northeast China and in sites of ancient iron-making workshops in central China. These findings prove that coal was then widely used in daily life and production. Another geographical work of ancient China, *Shui Jing Zhu* (*Notes on Waterways Classic*), describes the Bingjingtai Coal Mine, built by noted statesman Cao Cao (155–220) in 210 in Yexian County (located in Henan Province today). The mine was 50 m underground, with an output of several thousand tons. In the Song Dynasty, coal mining experienced a boom, and several large coal mines were discovered. The Song government set up a special institution to take charge of coal mining, and monopolized the trading of coal. A contemporary writer noted that in the Song capital Bianliang "several million people use coal as fuel, and none of them burns firewood." In recent years archaeologists excavated a site of an ancient coal mine in Hebi, Henan Province, and found that the mine was of quite a high technical level, with fairly complete facilities. The mine had two shafts with a diameter of 2.5 m, and depth of about 50 m. There were two tunnels of about

Shan Hai Jing

An ancient classic of Chinese literature, *Shan Hai Jing* (*Classic of Mountains and Seas*) consists of eighteen stories, including five pieces about mountains and thirteen pieces about seas. Very little is known about the author and the age of these stories. It is estimated that most were written in the period from the Warring States Period to the Western Han Dynasty. The stories mainly introduce geographical knowledge in folktales, such as mountains and rivers, ethnic groups, products, medicinal herbs, sacrificial ceremonies, and witch doctors. Containing many ancient fairy tales and legends, the book casts light on the history, geography, culture, transportation, and folk customs in ancient China, including the first known record of minerals.

Shui Jing Zhu

This is a famous work about China's ancient geography, written by Li Daoyuan (466 or 472–527) of the Northern and Southern Dynasties. Consisting of more than 300,000 Chinese characters in 40 volumes, the book was written as a commentary on the ancient classic *Shui Jing* (*Waterways Classic*). Although based on the *Waterways Classic*, Li Daoyuan greatly expanded the ancient work based partly on his own travel experiences. He introduced 1,252 waterways, big and small, and in addition to these, he also recorded the mountains, cities, local conditions and customs, and other interesting matters. The *Shui Jing Zhu* (*Notes on Waterways Classic*) is not only a famous scientific work with substantial content, but also valuable literature because of its outstanding language.

Mining coal.

Portrait of Marco Polo.

500 m, with a height of 2 m and width of 2.1 m, and the coal cutting workface was 1.4 m wide at the top and 1 m at the bottom. Although rather narrow, it was enough to cope with the needs of coal mining. When the mining of coal was completed, the miners would retreat jumping from grid to grid as if playing hopscotch. The mine also had fairly good facilities and mechanisms for ventilation, lighting, shoring, step-by-step lifting, and drainage. According to a narration in Song Yingxing's Exploitation of the Works of Nature, they had also devised a method of dealing with gas, the major risk in coal mining. Before cutting coal, the miners would push a big bamboo cane, which was hollowed inside and cut sharp at one end, into the coal layer to let out the gas.

The Venetian traveler Marco Polo (c. 1254–1324) mentioned with curiosity in his travel account that the Chinese burned a kind of "black stone," which burned as easily as wood but had a much stronger flame, and that the fire would last until the next day. This suggests that the Western world might not have known about coal at that time, yet by then the Chinese had been using coal for a thousand years. In addition, Westerners did not know how to have lighting in coalmines, and for a long period they cut coal in the dark. They did not solve the problem of coalmine drainage until the seventeenth century, or the issues of gas and ventilation until the eighteenth century. As noted by one author, in European coalmines a spark would turn the mine into a giant cannon tube, the heat waves raging through every tunnel and gushing out of the shaft with fragments of rock.

Pottery and Porcelain

Pottery is an important indicator of the New Stone Age, and the making of pottery ranks, with the use of fire, as a major human achievement. To make pottery, clay is first shaped and then hardened by drying and firing. Almost all nations have independently developed their own pottery techniques. When ancient humans settled down and began taking up farming and animal husbandry, pottery utensils became necessities for cooking, eating, drinking, and storage. In China, pottery dates back 8,000 years, and Chinese pottery and porcelain has been influenced by the economy, culture, science, and technology. In the long history of the nation, the Chinese made pottery and porcelain goods and explored and appreciated the art of ceramics, which became an inalienable part of Chinese civilization.

Early earthenware developed gradually. Early examples were mainly gray, black, and white, then some were impressed with lines and figures, and later some were decorated with colorful, opaque glazes. The shapes of earthenware also evolved from simple to complex and from coarse to delicate. The best known of the early works of pottery from ancient China are the terracotta warriors and horses excavated in Lintong District, Xi'an, Shaanxi Province, from the site of the tomb of Emperor Shihuang of the Qin Dynasty. The terracotta army was discovered accidentally in 1974. Since then more than 6,000 life-size terracotta warriors, more than 30 ceramic horses and over 1,000 chariots have been unearthed from three rectangular pits, which were arranged in a shape that looked like the Chinese character "品". Now a modern structure covers the three pits, and they constitute the on-site Museum of Qin Terracotta Warriors and Horses. Pit No.1 is the largest: 230 m east to west, 62 m north to south, and 5 m deep. It totals 14,260 sq m in area and houses an army formation. The warriors include soldiers, horsemen, armored soldiers, officers, and generals, some standing, and some kneeling, with some horses among them. The smaller Pit No. 2 houses a cavalry formation, with a large number of chariots. The smallest is Pit

Painted pottery urns of the Majiayao Culture, about 3000–2000 BC.

Army formation in Pit No. 1, Museum of Qin Terracotta Warriors.

No. 3, which houses 60–70 officers and generals, apparently a headquarters. The warriors and horses were skillfully made. The warriors look strong and dignified, with vivid hair, beards, individual facial expressions, and clothes, the latter with folds and wrinkles. The horses are strong and seem ready to go into battle. The figures were originally painted with bright mineral pigments, most of which have now disintegrated. These warriors are funeral objects, made to defend the Emperor in the next world. The Qin people must have paid a great deal to build such a terracotta army, but it also left us with a priceless treasure. Jacques Chirac, the former president of France, said in 1978 when he visited the Museum: "One can't claim to have visited China unless one has seen these terracotta warriors."

Tang Tricolor marked the second heyday of pottery in ancient China. The pottery artworks of the Tang Dynasty were decorated with three glazes—yellow, green, and white, hence the name Tang Tricolor. This kind of pottery was developed from the green and brown-glazed pottery of the Han Dynasty, and they were quite well known at home and abroad at the time they were made. Tang Tricolor works include horses, camels, beauties, dragon-decorated cups, and musician figurines. Among these the tri-colored horses are the most common, while the tri-colored camels are some of finest. The camels, usually carrying silk or musicians on their backs, were led by red-bearded, blue-eyed Tartars from central Asia, who wore narrow-sleeved gowns, and hats with upturned brims. These figures are reminiscent of scenes on the ancient Silk Road. Tri-colored pottery works have been in production for more than 1,300 years, and they have incorporated features of other art forms, such as traditional Chinese painting and sculpture. The colored glazes applied on the pottery flowed and seeped at high temperatures in the kiln, producing natural shading on the final artworks. They constitute a landmark in the history of Chinese pottery.

The third heyday of Chinese pottery was marked by *Zisha Tao*, or purple-clay pottery. Made using purple clay with high iron content and fired at a temperature of 1,200 degrees Centigrade, such unglazed earthenware shows different shades of purple. Most *Zisha* pottery wares are shaped through cementing pottery shards molded from *Zisha* clay. High-quality *Zisha* pottery, mainly teapots, are elegantly shaped by hand by skillful potters, while popular goods for large volume production are made by slip casting. *Zisha* wares were first mentioned in Song Dynasty classics, and *Zisha* pottery reached its height in the Ming and Qing Dynasties. The most celebrated *Zisha* wares are produced in Yixing, Jiangsu Province, and the finest are sold at astonishingly high prices. Although unglazed, *Zisha* pottery is waterproof. *Zisha* teapots can retain the fragrance of tea even until the following day, and they are also heat-resistant.

Tricolor camel with musicians. Tang Dynasty, unearthed from a tomb in Xian, Shaanxi Province.

China is universally acknowledged as the homeland of porcelain, reflected in its alternative name of china. Ceramics include both earthenware and porcelain, and their differences lie mainly in the clays used to make them. Earthenware or porous pottery is made of porous clay, while porcelain uses porcelain clay or kaolin. Porcelain clay contains kaolinite, quartzite, feldspar, and mullite, with little iron content. Porcelains

Ming Dynasty *Zisha* pot. The finest *Zisha* pots are usually plain.

Porcelain wine vessel. Song Dynasty, made in Jingdezhen.

Underglaze blue porcelain.

are fired at 1,200 to 1,300 degrees, and both the body and the glaze applied on it are fired at such high temperatures. The final product is white, rigid, and with little or no water absorption. When struck, it sounds rather metallic. "Celadon" is a term used for a particular type of porcelain, and also for its characteristic green glaze. Celadon appeared after the development of lead glaze and became a competitor with blue and white porcelains. In the Song Dynasty, Chinese porcelain saw its first peak of development with six major varieties made in the kilns of Ding, Yaozhou, Jun, Cizhou, Longquan, and Jingdezhen. These were all ordinary kilns that met the needs of all sectors of society. There were also Imperial kilns that provided wares for the Imperial court: Ru, Ge, Ding, Jun, and Guan.

Celadon wares from the Yuan Dynasty were marked by their large size and weight. The decorations were painted delicately in bright blue, and covered by *Yingqing* glaze. These majestic wares were mainly high-end products from the Fuliang Porcelain Factory. There are also some smaller celadon wares from the Yuan Dynasty. The glaze used on such wares is of an egg-white color, and their decorations are simple and straightforward. Such wares were made in ordinary kilns for daily use. The decorations are mainly pine, plum, bamboo, lotus, dragon, phoenix, crane, deer, people, flowers, birds, and grass. Some of them also feature historical stories. Large, high-quality celadon wares from the Yuan Dynasty are extremely rare today—they number only several hundred, and are mostly held in foreign collections because they were made mainly for export. In

Porcelain kiln. Chinese kilns have been in production for 1000 years.

the 1950s, Chinese porcelain experts learned that there were a large number of Yuan celadon wares kept in a museum in Istanbul, Turkey, and they were eager to view them. But the Turkish authorities were so cautious that the visit was delayed time and again. It was not until the early twenty-first century that the hopes of Chinese experts came true, and then only after repeated consultations, diplomatic maneuvers, and approval by the Turkish Prime Minister. For three consecutive days, seven Chinese experts examined the treasured wares under the supervision of Turkish armed guards. The experts were greatly excited by the forty Yuan celadon wares, all large and fully decorated. An elderly man among them recalled: "When I first saw the wares, I sobbed softly. I touched them again and again in admiration." Another elderly man said: "I have realized a lifelong dream to see the treasured wares. Now I will have no regret when I die."

In the Ming and Qing dynasties, Chinese craftsmen in the porcelain industry carried on the fine traditions of previous empires, and the industry moved to new heights with Jingdezhen as its center. People praised a variety of porcelain from Jingdezhen, saying it is: "as thin as paper, as bright as mirror, as white as snow, and as resonant as chime stone." This suggests the perfection of Ming and Qing porcelains.

Wine and Liquor

The ancient Chinese were the first to discover wine and to invent the technique for making wine and liquor. The earliest wine in the world was produced when grain, fruit, and milk fermented under certain natural conditions. An ancient writer noted that monkeys in the mountains collected flowers and fruits and made wine for themselves. Though by accident, it was only natural for ancient people to discover the existence of wine; but the making of wine was indeed a great invention. *Lüshi Chunqiu* (*Lü's Spring and Autumn Annals*), which was compiled in the second century BC, writes that "Yi Di was the first to make wine." A book written in the Han Dynasty also states: "Yi Di was the first to make wine, and Shao Kang developed a fine wine." Judging from the wine vessels unearthed in archaeological sites, wine making probably started in the period of the Yangshao Culture, about 4,000 to 5,000 years ago. Very likely the technique was

Poetry and wine have always been linked in Chinese culture. Without wine, there would be no Li Bai, the Tang Dynasty "drinker-poet."

Yeast manufacture in ancient China.

developed not by any one master but by a group of people. However, people used to attribute it to several masters, the most well-known being Du Kang.

Because of natural and social conditions, it was not possible to make wine from wild fruits on a large scale. In China, a large grain producer, large-scale production of wine was possible only when grain was used as the main ingredient. Grain, though full of starch, cannot be directly fermented to turn into wine. The starch must first be saccharified to become glucose, after which the glucose may be fermented to become alcohol. The agent that helps to turn starch into glucose is a primitive yeast, which grows when cereals mold and bud under warm, wet conditions. After a short period of saccharification, the cereals can produce a light sweet wine. From the primitive yeast, a distiller's yeast was developed, which contains rich amounts of microorganisms that are useful in making wine and liquor. The primitive yeast was later abandoned as it could only produce a light wine. According to earliest Chinese classics, the primitive yeast and distiller's yeast were first used in the sixth or seventh centuries BC.

Brewers and distillers in ancient China developed various yeasts and yeast-making techniques, including powder and cakes. Jia Sixie, a historian of science in the Northern Wei Dynasty (386–534), recorded thirteen methods for making yeasts in his *Qi Min Yao Shu* (*Important Arts for the People's Welfare*). Using different yeasts and medicinal herbs, brewers and distillers turned out a range of spirits with various flavors.

The discovery and application of *hongqu* (red yeast) in the Northern Song Dynasty marked a major achievement in wine and liquor making in China. *Hongqu* is a major yeast for making liquor in the southeastern provinces of China, and is also used to make vinegar and preserved bean curd. Red yeast is also a high-quality condiment and natural coloring for foods. The major microorganisms in *hongqu*, genus *Monascus*, grow very slowly

Detail from *Han Xizai Ye Yan Tu* (*Han Xizai's Night Banquet*) by Gu Hongzhong, Five Dynasties.

under natural conditions, and brewers and distillers in ancient China developed various ways to accelerate the growth of the yeast. *Tiangong Kaiwu* (*The Exploitation of the Works of Nature*) tells how the growth can be sped up: select fine molds as the seeds, add alum to increase acidity and control the growth of unnecessary molds, and add water by steps to promote the growth of the yeast. These advanced techniques for making yeasts boosted the wine and liquor industry. The historian Jia Sixie recorded more than forty methods for making wine and liquor. As a land of brewers and distillers, China now turns out the well-known liquors of *Moutai* and *Wuliangye*, and *Changyu* wine.

Sericulture

Silk fiber is an important material for clothing, and it was the ancient Chinese who developed this valuable material for textiles.

Sericulture is the most successful example of exploiting insect resources. Legend has it that Leizu, wife of Huangdi (Yellow Emperor), was the first to teach women to pluck mulberry leaves and feed silkworms. People learned to obtain silk filaments from wild silkworms before they were domesticated, and domestication of silkworms began about 5,000 years ago. In a New Stone Age site in Shanxi Province in northern China, archaeologists found half a cocoon, which was cut more than 5,000 years ago. In another site from the same period in Zhejiang Province in eastern China, archaeologists unearthed silk fabric, strips, and thread, dating from 4,700 years ago. Inscriptions on Shang Dynasty bones or tortoise shells included the characters for *can* (silkworm), *sang* (mulberry), *si* (silk), and *bo* (silks), along with records of sacrifices to the god of mulberry trees and inspections of silkworm production. This indicates that by then sericulture had become part of the daily life.

Sericulture was quite well developed in northern China. A poem about sericulture in Shaanxi Province in *Shi Jing* (*The Book of Songs*) reads: "In a shiny day in spring, orioles are singing. On a small path, women carrying baskets are going to pluck tender mulberry leaves."

Spinning Wheel, by Wang Juzheng, Northern Song Dynasty.

The *Picture of Picking Mulberry Leaves* on a bronze ware item of the Warring States Period vividly depicts women collecting mulberry leaves. Another poem in *The Book of Songs* reads: "Among their ten-acres, the mulberry-planters stand idly about," while a chapter in *Mencius* says: "To plant mulberry trees in a five-*mu* compound, will turn out enough silk to clothe fifty people." Xun Kuang (c. 313–230 BC), a noted thinker of the Warring States Period, wrote in *Ode to the Silkworm* that after three periods of dormancy, the silkworm builds its cocoon. The *Li Ji* (The Book of Rites), compiled 2,000 years ago, records methods for preventing silkworm diseases:

Spinning wheel, depicted in a pictorial carving.

Gathering mulberry leaves.

Raising silkworms.

Colorful silk fabrics added to the dignity of monarchs and their officials, and to the beauty of women.

to wash silkworm eggs in vermilion solution, salt water, or limewater.

Raising silkworms contributed directly to the development of silk weaving, which remained a unique Chinese technique for a long time. Colorful silk fabrics added to the dignity of monarchs and their officials, and to the beauty of women. After the opening of the Silk Road, fabrics became a major export of ancient China. For a thousand years the several-thousand-kilometer road served as a major channel of exchanges between China and the rest of the world, and a driving force for economic and cultural development in China. All silkworm-raising countries in the world obtained their silkworm eggs and sericulture techniques from China, either directly or indirectly: Korea, 3,000 years ago; Japan and Vietnam, 2,000 years ago; central Asian countries, 1,600 years ago; Europe, 1,400 years ago; and South America, 400 years ago.

Zhang Qian and the Silk Road

In 138 BC, Emperor Wudi of the Western Han dispatched Zhang Qian (164–114 BC) on a diplomatic mission to Dayuezhi, one of the states in the Western Region – the region west of Yumenguan Gate and Yangguan Gate, in Gansu Province. His mission was to make a joint effort to fight against Xiongnu (the Hun). But Zhang Qian was arrested en route and detained by Xiongnu for a period of 10 years. Later, he managed to escape, and reached Dayuan (Ferghana), Kangju (Kangar), Dayuezhi and Daxia (Bactria). On his way back, he was again detained by Xiongnu for one year. In 126 BC, he fled following internal disorder of Xiongnu and came back to Chang'an, the capital of the Western Han. In 119 BC, Zhang Qian was sent to the Western Region again and enabled the Western Han to establish friendly relations with states such as Wusun. Zhang Qian finally completed the epic mission of exploring Central Asia and pioneered the Silk Road connecting China to the West.

Tea and Tea Culture

Tea, coffee, and cocoa are three global beverages. The ancient Chinese were the first to cultivate and process tea and prepare the drink, and the product remains a major export.

The tea plant is native to southwest China. A Chinese classic writes: "Tea plants are native of the south. Some of them are one to two feet high or several dozen feet high, and in the Bashan Mountains and the Yangtze Valley some tea trees are so big that it takes two persons to stretch their arms to circumvent the trunk." According to legend, it was Shennong, the god of farming, who discovered the tea plant. Shennong was in the process of tasting all kinds of herbs when he ingested some toxic plant and fell into a coma. When he started to regain consciousness, he took some leaves from a plant and chewed them. To his surprise, he recovered fully and felt energetic. He then plucked the leaves to treat people who fell ill, thus beginning the use of tea. Lu Yu (733–c. 804), a tea expert of the Tang Dynasty, writes in his *Classic of Tea*: "It was Shennong who began using tea as a beverage."

The cultivation of tea involved the use of tea species of short shrubby growth, rather than trees. *Er Ya*, the earliest Chinese dictionary, includes the character for tea, and states that tea tastes bitter. When King Wuwang of the Zhou Dynasty fought the last

Tea Party, by Wang Chengpei, Qing Dynasty.

Taking tea.

emperor of the Shang Dynasty and conquered the area of today's Sichuan, Wuwang ordered the local vassals to send lacquer, honey and tea as tributes. In the Qin Dynasty, tea plants spread to Shaanxi, Gansu and Henan, but tea remained a precious product, not available to ordinary people. In the Eastern Han Dynasty Buddhism was introduced to China, and the consumption of tea gradually increased, for the refreshing beverage was good for monks who sat for days on end reading Buddhist scriptures, and the land around Buddhist temples in the mountains was good for growing tea plants. Tea plantations were developed in the Tiantai and Emei Mountains near Buddhist temples, and with increasing tea output the beverage gradually spread among ordinary people.

In the 500 years between the Eastern Han and the Northern and Southern Dynasties (420–589), tea plants were introduced to the Huaihe Valley, to the middle and lower reaches of the Yangtze River, and to areas south of the Five Ridges, where conditions were ideal for the growth of tea plants. By the Tang Dynasty, tea had become a popular beverage. Two writers note in their essays: "Near the capital there are many tea shops in cities, where customers pay money and get the beverage themselves," and even in the north: "people can do without food for several days but not for a single day without tea." By then tea was produced in fifty states and prefectures, roughly equivalent to fifteen provinces in central, east, south, and southwest China today. During the reign of Emperor Dezong of the Tang Dynasty, tax on tea amounted to 400,000 *min* a year, and the output value of tea production was about 4 million *min* (one *min* equals to 1,000 copper coins).

Ancient tea farmers accumulated rich experience in tea production, including the selection of the sites for tea plantations, encouraging budding, irrigation, fertilization, and shading for tea trees. They also exchanged tea for animal products from nomadic people. Attracted by the flavors of the beverage, men of letters developed a tea culture. From the Tang to the Yuan and Ming Dynasties, tea farmers developed various methods for processing tea leaves and buds in different localities. Classification of teas

Preparing Tea, painted by Ren Xiong, Qing Dynasty.

Wooden statue of Lu Yu, the Sage of Tea; Tea Museum in Hangzhou, Zhejiang Province.

by the manufacturing process results in the three categories of tea: fermented (black), unfermented (green), and semifermented (oolong or pouchong). The prevalence of tea drinking also led to the improvement of tea-processing devices and tea sets. Since the Ming and Qing Dynasties, tea has been a major export from China.

Lu Yu, the tea expert of the Tang Dynasty, wrote *Classic of Tea*, the world's earliest work devoted to tea culture. He discussed the origin of tea, its cultivation, picking of tea leaves, processing, and preparation of the beverage, the quality of water, and tea sets, as well as anecdotes about tea. This encyclopedic work was so impressive that people praised its author as the Sage of Tea.

Tea was introduced to other Asian countries in the fifth century, to European and American countries in the seventeenth century, and the beverage gradually prevailed throughout the world. China not only exported tea products, but also provided tea plants and seeds for other countries. Nowadays tea plants are grown in many countries, including Japan, India, Sri Lanka, Indonesia, and Russia. The Japanese even developed *sado*, a tea ceremony, based on Chinese tea culture.

Savoring tea has a long tradition in Chinese culture.

Other Major Inventions (II)

Star Catalogue and Star Atlas

Astronomy emerged quite early in ancient China. Some relics from the Neolithic period featured astronomical signs and symbols. Sima Qian (c. 145 or 135 BC–?), the great historian of the Western Han Dynasty, writes in his *Shi Ji* (*Historical Records*): "The Yellow Emperor made observations of stars and worked out a calendar. He established the five elements of metal, wood, water, fire, and earth, and understood their correlations, and he solved the problems of leap months and leap days in the calendar. He divided all things on earth under the heaven and the God into five groups, and brought about their orderly operations, free from disturbances and confusion. Therefore the people believed in his rule, and the God praised his wisdom. He showed high respect to the people and the God, though they were quite different. As a result the people had the blessing of the heaven, and enjoyed an affluent life."

This sowed the seed of the great concept of "integration of nature and man" and formed the ideological basis for astronomical undertakings. Soon after the Yellow Emperor, a post of *Huozheng* was established by the Imperial court to take charge of astronomical and calendrical affairs, making close observations of the planet Mars, so as to direct farming activities according to the positions of the planet. The duties of ancient officials in charge of astronomical affairs were to observe the movements of the sun, the moon and stars, to forecast solar and lunar eclipses, determine the times of the solar terms, and work out

Pottery vase with petal-like sun patterns, from a site the Dawenkou Culture, 6,000 years ago.

Chinese constellations
To facilitate star observation, the people of ancient China divided stars into different groups and named them after earthly objects. A group of stars was called a *guan*, or constellation, also known as *zuo* after the Tang and Song Dynasties. These groupings are not the same as the constellations we recognize today.

calendars. With diligence, wisdom and perseverance, they made great contributions to astronomical observations and formulation of calendars in ancient China.

Gan De from the state of Qi and Shi Shen from the state of Wei were both royal astronomers. Shi Shen was the author of *Tianwen* (*Astronomy*) in eight volumes, and Gan De also wrote an eight-volume *Tianwen Xingzhan* (*Astronomical Observations of Stars*). But both of their works have long been lost. In later years other writers collected excerpts from the two works that were quoted in other books, and compiled these into a book known as *Gan Shi Xing Jing* (*Gan's and Shi's Classics on Stars*). Both of them recorded a number of fixed stars, including their names and positions, but some of them overlapped. In the Three Kingdoms Period astronomer Chen Zhuo listed all the fixed stars recorded by Gan De, Shi Shen, and another astronomer Wu Xian, and divided them into

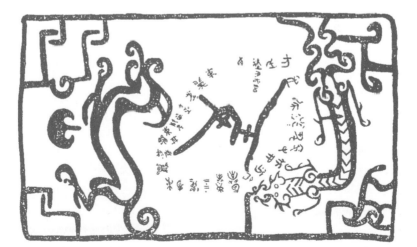

The twenty-eight lunar constellations.

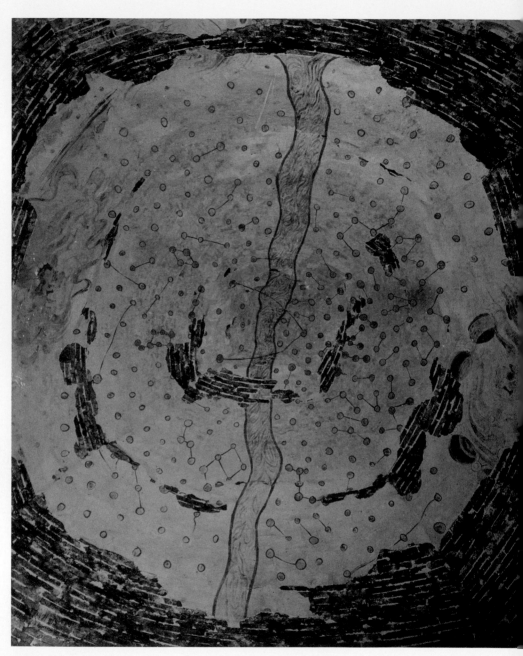

Northern Wei Dynasty star map from a tomb, unearthed in the suburbs of Luoyang, Henan Province in 1974. Notable is the depiction of the Milky Way, rarely seen in other star maps.

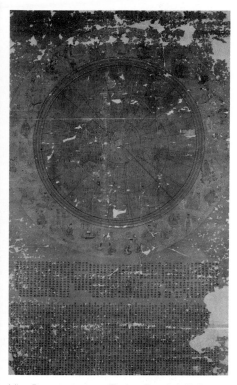

Ming Dynasty star map, Tianhou Temple in Putian, Fujian Province. This map is a useful source for studying ancient navigation using stars.

283 groups. All the 283 *guan* (Chinese constellations), with 1,464 fixed stars, were marked in different colors on a star map. Later astronomers used the map to work out a star atlas and a celestial sphere. In *Gan's and Shi's Classics on Stars*, 120 constellations with 815 stars were attributed to Shi Shen, and 146 constellations with 687 stars to Gan De.

What is most valuable is that Shi Shen provided values of the coordinates of the 120 standard stars in 120 groups of stars, including their equatorial coordinates. The coordinates of the twenty-eight lunar constellations were expressed in the ascensional differences of their representative stars as arranged along the equator from west to east, and the angle between the fixed star and the celestial pole. As for other fixed stars in the constellations, their coordinates were the ascensional differences between them and the representative star, and the angles between them and the celestial pole. This shows Shi Shen's star catalogue adopted the equatorial coordinate system, or the right ascension and the declination to show the position of any celestial body on the celestial sphere. Shi Shen's star catalogue was expressed completely in numerals. The adoption of the equatorial coordinate system was a unique contribution of the ancient Chinese to world astronomy, as ancient astronomers in the West all used the ecliptic coordinate system to mark the positions of fixed stars.

Astronomers in ancient Greece, from the time of China's Warring States Period until the sixteenth to seventeenth centuries, used the ecliptic coordinate system inherited from Babylon. The combination of the equatorial coordinate system and the ecliptic coordinate system helps to improve the accuracy of astronomical observations and records. The astronomical achievements of Shi Shen and Gan De were all recorded in *Kaiyuan Xingzhan* (*The Kaiyuan Star Observations*) from the Tang Dynasty. Although Shi Shen's star observations have been improved in the several hundred years since his time, his star catalogue is nevertheless one of the earliest in the world, and has high scientific value. Quite a few figures in the star catalogue were the results of observations made in the Warring States Period, which suggested that Shi Shen had used angular surveying devices to determine the positions of celestial bodies in the equatorial coordinate system. This in turn shows that Chinese astronomers in the Warring States Period led the world in astronomy and in the making of astronomical instruments. Shi Shen and Gan De were involved in another unique achievement: they divided the celestial sphere into 365¼ degrees. As a tropical year has 365¼ days, each day the sun moves one degree on the celestial sphere. Yet in the West, most ancient astronomers adopted the Babylon division, dividing the celestial sphere into 360 degrees. Shi and Gan also made fairly accurate observations of the five major planets. Both of them noted that the sidereal orbit period of Jupiter was 12 years (compared with the exact value of 11.86 years). Shi Shen also found through observations that the synodic period of Jupiter was 400 days (compared with the exact value of 398.9 days), the synodic period of Venus was 587.25 days (compared with the exact value of 583.9 days), and the synodic period of Mercury was 136 days (compared with the exact value of 115.9 days).

Observation of Solar and Lunar Eclipses

Records of solar eclipses, engraved on an ox bone. Shang Dynasty, unearthed in Anyang, Henan Province.

All dynasties in ancient China carried a fine tradition of recording in detail the results of astronomical observations. From the Han to the Yuan Dynasties, 596 solar eclipses were recorded; the ancient Chinese recorded 2,000 lunar eclipses, including 400 total eclipses. Outside China the earliest records of solar eclipses were found in the ruins of ancient Babylon, and the earliest solar eclipse among the six records took place in 911 BC. Yet in China the earliest record of a solar eclipse, found in the inscriptions on bones and tortoise shells unearthed at the Yin ruins, was in 1200 BC, nearly 300 years earlier than the ancient Babylon record, and more than 600 years earlier than the first record of a solar eclipse in Europe. *The Spring and Autumn Annals* recorded 37 solar eclipses in 244 years, and 32 of these have been proved reliable. Ancient Chinese observers made detailed records of the eclipses, including the first contact, the maximum phase, and the last contact. Ancient Chinese astronomers were also able to work out the cycle of repeated solar eclipses to be 777 nodical months (solar months), or 716 lunations (lunar months). This doubles

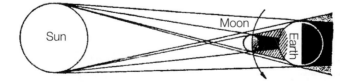

Diagram showing total and partial solar eclipses.

Solar eclipse.

the cycle of 388.5 months computed by American astronomer Simon Newcomb (1835–1909), but Newcomb calculated the cycle more than 1,000 years later. As their forecasts were so accurate, ancient Chinese astronomers were able to tell people to prepare a basin of water on the date a solar eclipse was due to occur so that they could observe the whole eclipse process, with the image of the sun reflected in the water. By studying ancient Chinese records of eclipses and other astronomical phenomena, Western scientists have established a new branch of astronomy—historical astronomy.

Astronomers in ancient China also made close observations and detailed records of lunar eclipses. China's earliest record of a lunar eclipse is found in *the Book of Songs*, the first collection of ancient Chinese poetry. A verse in the book says: "The moon was eclipsed, and it is a normal celestial phenomenon." Studies found that this lunar eclipse occurred in the eighth month of 776 BC. It is also the earliest known record of a lunar eclipse, 55 years earlier than the record in ancient Egypt of a lunar eclipse that occurred in the second month of 721 BC. Astronomers in ancient China also explored the cause of lunar eclipses. *The Book of Changes* says a lunar eclipse usually comes after increscence, or around the fifteenth day of the lunar month. Zhang Heng (78–139), a great scientist of the Eastern Han Dynasty, gave a much clearer explanation for lunar eclipses. He said that since the moon is illuminated by the sun, a lunar eclipse took place when the earth blocked the sunlight.

Surveying of the Meridian

Astronomers in ancient China established the concept of the meridian quite early. They divided the meridian into 90 degrees from the North Pole to the equator, and they knew that the shadow under the sun decreased from north to south. However, since they had not formed a clear idea that the earth is spherical and had not made field surveys, they did not know the exact proportions of shadows in relation to their geographical locations. For a long time they believed that there was a difference of one *cun* (roughly an inch) in the length of shadow in relation to a distance of 1,000 *li* (about 500 km) from north to south. During the early years (about 604–607) of the reign of Emperor Yangdi of the Sui Dynasty, astronomer Liu Zhuo (544–608) suggested that a surveyor with mathematical knowledge be invited to conduct surveys on the plains north and south of the Yellow River, from north to south, and measure the distances of several hundred *li* with ropes. He said this method could be adopted in studying the earth, the sky, and stars. In 607, Emperor Yangdi ordered that various localities measure the length of shadows, but the project failed because of Liu Zhuo's death. More than 100 years later, the monk Yixing (683–727), an astronomer of the Tang Dynasty, carried out the historic mission of measuring the meridian on the ground.

Yixing was also known by the name of Zhang Sui before he became a monk. A native of Changle, Weizhou (now Nanle County, Henan Province), Zhang Sui studied diligently in his childhood and grew to become a learned man. Zhang refused to serve Wu Sansi, a corrupt, powerful and tyrannical minister, and he secluded himself in a temple in Songshan Mountain and became a monk. He was thus known as the Monk Yixing. In 717, Emperor Xuanzong of the Tang Dynasty invited Yixing to Chang'an, the capital of the empire, to serve as an astronomy advisor. There Yixing compiled and popularized

Monk Yixing.

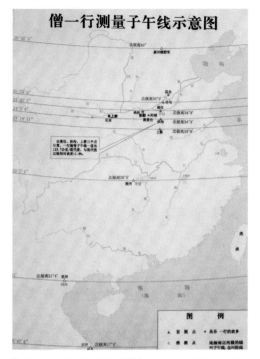

僧一行测量子午线示意图

Sketch map showing Monk Yixing's survey of the meridian.

the Dayan Calendar, extended the application of Liu Zhuo's formula on the variable-speed motion of the sun, and, together with another astronomer Liang Lingzan, created a number of astronomical devices including a bronze armillary sphere and an ecliptic armillary sphere. With these new devices, Yixing determined the positions of more than 150 stars, and made repeat measurements of the degrees of the twenty-eight lunar constellations from the North Pole of the celestial sphere. His observations proved that a number of previous records had been inaccurate.

Based on his own observations, Yixing deduced that the stars were constantly changing their positions in the celestial sphere. He thus became the first astronomer in the world to study the movement of stars, about one thousand years before English astronomer Edmund Halley (1656–1742).

As major errors had been found in the forecasts of solar eclipses based on previous calendars, Emperor Xuanzong assigned Yixing to the task of making a better calendar. Yixing decided to base his compilation on field surveys, and in 724 he launched and directed a giant project to measure the meridian through field surveys. The sites selected for the surveys included Linyi (in the centre of present-day Vietnam, at about latitude 18° North), in the south, and Tiele (in today's Mongolia, at latitude 51° North) in the north. In between were sites in Annam (in today's Vietnam), Wuling,

Langzhou (now Changde, Hunan Province), Xiangzhou (now Xiangfan, Hubei Province), Wujinguan, Shangcai in Caizhou (now Runan, Henan Province), Fugou, Xuzhou (now Fugou, Henan Province), Taiyuetai, Junyi in Bianzhou (now Junxian, Henan Province), Baima, Huazhou (now Huaxian, Henan Province), Taiyuan Fu (now Taiyuan, Shanxi Province), Hengyejun, Weizhou (now Weixian, Hebei Province), Yangcheng (now Gaocheng Town, Dengfeng, Henan Province), and Luoyang (now Luoyang, Henan Province). Of all the results, those by Nangong Yue and others at Baima, Junyi, Fugou, and Wujin were the most satisfactory. At these sites, which ran for several hundred *li* from north to south, the surveyors recorded the differences in the length of shadows at the summer and winter solstices, the vernal and autumn equinoxes, measured the distances of the four sites, and surveyed the heights of the North Star (latitudes) at the four sites. By calculating the measurement results they found that for one degree of height of the North Star (latitude) the corresponding distance on the ground was 351 *li* and 80 *bu*, or 129.22 km. This is 18.02 km longer than the present-day value of 111.2 km, which is the length of one degree of arc of the meridian. The Monk Yixing thus overthrew the erroneous concept that one *cun* in the shadow corresponded to 1,000 *li* in distance. Joseph Needham believed this was pioneering work in the history of astronomy, and occurred ninety years earlier than the findings of Al Mamun, the ancient Arab astronomer, in the Euphrates Valley.

Zhang Heng and his Seismograph

Astronomers in ancient China created a large number of devices for astronomical observations, and many of these were able to show clearly the changes in the positions of celestial bodies. The celestial equator is a great circle on the celestial sphere in the same plane as the earth's equator, while the ecliptic

Zhang Heng.

is the apparent great-circle annual path of the sun in the celestial sphere, as seen from the earth. The ecliptic plane intersects with the celestial equator at a deflection of 23.26°. The sun meets the celestial equator on the two solar events of the vernal and autumnal equinoxes. The devices made by ancient Chinese astronomers were mainly used to show the difference between the equatorial system and the ecliptic system, and their conversion. Zhang Heng (78–139), a noted astronomer of the Eastern Han Dynasty, was an expert at making such devices. One of the devices made by Zhang Heng is the *hun xiang*, a kind of celestial globe. In the *hun xiang*, water from a "clepsydra" (water clock) drives a number of gears so that the globe turns on its axis in tempo with the movement of the earth. The changes in the celestial sphere are thus shown accurately and vividly. Before that, astronomers in ancient China had created other similar devices, such as *hun tian yi, jian yi, yang yi*, and *gui yi*. But all these devices, including the *hun xiang*, are quite complicated to explain in simple words. In comparison, the principle of the *di dong yi* or seismograph invented by Zhang Heng might be easier to understand.

Shocked by the catastrophes caused by earthquakes, Zhang Heng set out to create a device that could both report earthquakes that occurred in remote places and forecast major earthquakes by detecting preceding minor tremors. Legend has it that Zhang Heng was once riding in a wagon, and his driver stopped the horses abruptly in an emergency, upon which Zhang was almost thrown out of the wagon by inertia.

Model of the *di dong yi* (seismograph).

This experience inspired him: as an earthquake took place in an instant, an object would be shocked by its inertia; if this force was recorded, would it not be possible to report and forecast earthquakes?

The seismograph invented by Zhang Heng had a huge copper body—about eight feet in diameter—roughly the shape of a barrel, and on the barrel were eight dragons with their heads pointing downward, set at eight directions: east, west, north, south, northeast, southeast, northwest, and southwest. In the mouth of each dragon was a copper ball, and right under each dragon's head was a copper toad with its mouth wide open. Inside the barrel was a complicated mechanism, of which historical records gave few details. Researchers believe there must have been a suspended pendulum connected with eight levers controlling the mouths of the dragons. At normal times the whole mechanism was in a state of unstable equilibrium. When a quake took place, the tremor wave would break the balance, and a lever in the direction of the quake would be activated to open the mouth of a dragon. The copper ball inside the dragon's mouth would then drop into the mouth of the toad below, and observers would then hear the sound and know that a quake had occurred. The device was so well designed that only one dragon's mouth would open when a quake took place, while the remaining seven remained shut.

The seismograph was set up in Luoyang, the capital of Eastern Han Empire, where it reported accurately several earthquakes. However in 138, the eighth year of the reign of Emperor Shundi, something strange took place. One day, the dragon that faced west opened its mouth, and the copper ball dropped into the toad's mouth, but no one in the capital felt any tremor. However, several days later reports came from L'ongxi, several hundred miles west of the capital, saying there had been an earthquake. This event amplified people's admiration of the magic device. From then on, historians were ordered to record earthquakes in accordance with the reactions of the seismograph. Outside China, a similar device was made in Persia in the thirteenth century, more than 1,000 years after Zhang Heng invented his seismograph. The principle of Zhang's invention is still used today in modern seismographs.

The *di dong yi* was so impressive that it has become a symbol of the achievements of science and technology in ancient China, and its inventor Zhang Heng, has been remembered from generation to generation as a great astronomer and writer.

Guo Shoujing and his *Shoushi Calendar*

The ancient Chinese had long led the world in making calendars. According to legend, the Yellow Emperor, ancestor of the Chinese nation, worked out the first calendar in ancient China. *Shang Shu*, or *Documents of Ancient Times*, a book said to be compiled by Confucius, states that in the times of Emperor Yao a year was divided into 366 days and an additional month was established to decide the four seasons. In the Spring and Autumn Period, Chinese astronomers were the first in the world to intercalate seven months in nineteen years, to reconcile the differences between the lunar and solar calendars. In the late Spring and Autumn Period, Chinese

Guo Shuojing.

astronomers worked out the *Sifen Calendar*, which decided that the duration between two winter solstices was 365.25 days, the most accurate number of days for the tropical year. The number of days was the same as that of the Julian Calendar, but the Roman counterpart came 500 years later than the Chinese one. In 104 BC, Emperor Wudi of the Han Dynasty promulgated the *Taichu Calendar*, which set the tropical year at 365 385/1539 days and the lunar month at 29 43/81 days, and established for the first time the Twenty-four Solar Terms in the calendar. For more than 1,000 years after the *Taichu Calendar*, some seventy other calendars were formulated in ancient China. The well-known among them include the *Linde Calendar* (665) compiled by Li Chunfeng in the Tang Dynasty, and the *Dayan Calendar* by the Monk Yixing, but most of them were not used for long. The only exception was the *Shoushi Calendar* worked out by Guo Shoujing (1231–1316) in the Yuan Dynasty, which was used for more than 360 years, and was thus the most representative of ancient Chinese calendars.

In 1276, Kublai Khan, Emperor of the Yuan Dynasty, assigned the task of compiling a new calendar to the astronomer Guo Shoujing, so that his new empire would have a unified calendar from north to south, and the errors in previous calendars could be corrected. Guo was an exceptionally talented and dedicated scientist. On taking on the task, Guo said: "A good calendar must be based on observations, and observations depend on good instruments." He examined the *hun yi* (celestial globe), the only instrument at the observatory of the capital Dadu (now Beijing), and found that the North Star of it was set at 35 degrees, which was the latitude of Kaifeng, where the *hun yi* was made. This meant that the instrument had not been adjusted when it was transported to Dadu from Kaifeng. Also, after many years of war, the instrument had fallen into disrepair and could no longer be used. Guo thus made it a priority to develop new

Model of *jian yi* (simplified armillary sphere), invented by Guo Shoujing. This is a simplified version of *hun yi* (armillary sphere), using only two sets of rings. It was the most advanced device for astronomical observations at that time.

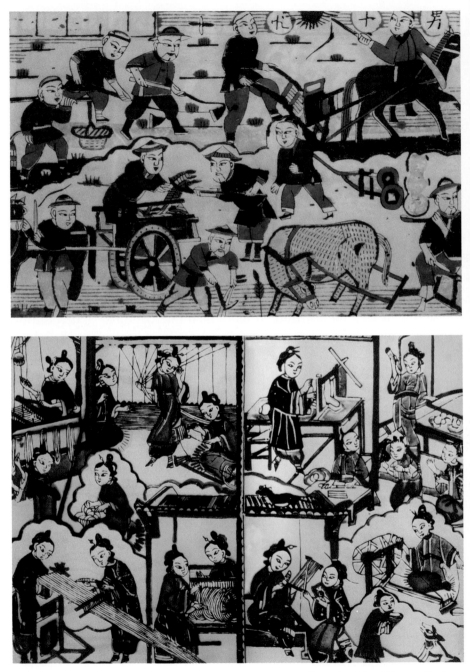

Men and Women at Work, a set of block-printed New Year illustrations. The tasks and activities are closely related to the seasonal changes of nature. The development of astronomy boosted production and living standards.

instruments. After three years of strenuous efforts, he designed twelve astronomical devices that were far better in function and accuracy than previous ones. He also made a number of portable instruments to be used in field studies outside Dadu.

As part of the calendar project, Guo presided over a nationwide program of astronomical observations. He selected twenty-seven sites for astronomical observation throughout the country, covering a wide area from latitude 15° North to latitude 65° North, and from longitude 128° East to longitude 102° East. Items of observation included the length of shadow of the gnomon (sundial blade), the angle of the North Star from the ground surface, and the beginning times of day and night on the vernal and autumnal equinoxes. All the records of the angle of the North Star from the ground had an average error of only 0.35. Guo also examined nearly 900 years of astronomical records—from 462 to 1278—and selected six figures from the records for calculating the duration of the tropical year. Guo's result was 365.2425 days, which was the same as that of the Gregorian Calendar (the calendar now widely used across the world), but came three centuries earlier. Pierre Simon de Laplace (1749–1827), a noted French astronomer, acknowledged that in the mid-thirteenth century Guo made the most accurate surveys of the length of shadow of the gnomon.

Guo Shoujing and other astronomers worked for four years and completed the calendar in 1280. Adopting the method of inscribing the circle by the sagitta of arc, they made numerous calculations converting the data of the ecliptic coordinate and the equatorial coordinate systems, and used twice interpolations to solve the variations in the speed of the sun's movement, which affected the accuracy of the calendar. The calendar was unprecedented in accuracy. It adopted the winter solstice of the year 1280, the ninth year of the Yuan Dynasty, as the epoch, the point of reference for the calendar, and established the duration of a tropical year at 365.2425 days, and that of a lunar month at 29.530593 days. The error between the duration of its tropical year and that of the revolution of the earth around the sun was only 26 seconds, as accurate as the Gregorian Calendar, which was invented 300 years later. The calendar was named *Shoushi*, meaning measuring time for the public.

The *Shoushi Calendar* soon spread to Japan and Korea, and was adopted in the two countries. In recent years, astronomers in the United States, Japan and other countries have shown renewed interest in this calendar, and organized translation and in-depth study of this work. This calendar has established its place in the history of world astronomy. To honor Guo, a crater on the Moon is named Guo Shoujing Crater, and a planetoid discovered in 1964 is named Guo Shoujing Planetoid.

The Decimal and Binary Systems

The decimal system is now universally used in mathematical calculations and the binary system plays a vital role in computer technology. These two systems are closely related to systems used in ancient Chinese exploration and innovations.

The decimal system, a positional numeral system employing 10 as the base and requiring 10 different numerals, looks quite simple and natural, but it took strenuous efforts to develop such a system. Counting of numerals appeared in China some 6,000 years ago in the late Neolithic Period. At that time people counted numbers by making knots on a rope and by cutting notches in wood. Numerals were found on pottery pieces unearthed at the Banpo site dating back 6,000 years, and on cultural relics dating from some 4,000 years ago that were unearthed in Shaanxi, Shandong, and Shanghai, there were single digits and signs representing 10, 20, and 30. This means that the decimal system was in use at that time. Of the thirteen characters used for counting found amongst

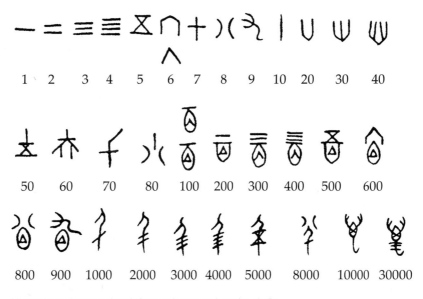

1	2	3	4	5	6	7	8	9	10	20	30	40

50	60	70	80	100	200	300	400	500	600

800	900	1000	2000	3000	4000	5000	8000	10000	30000

Numerals as shown on inscriptions on bones and tortoise shells.

the inscriptions on bones or tortoise shells of the Yin and Shang Dynasties, nine were identified as single digits, and the remaining four were symbols representing place values such as 10, 10^2 and 10^3. In the number system used in the inscriptions, numbers comprised the nine digits from one to nine and symbols of place values. Signs from the two groups were combined to represent certain place values, and groups of signs were then added together to represent numbers. In an inscription dated from the thirteenth century BC, "547 days" was written as "five hundred plus four decades plus seven days," and in *Yi Jing* (*Book of Changes*) the number "eleven thousand five hundred and twenty" was recorded. The way numbers were counted in the inscriptions on bones or tortoise shells has lasted through to modern times. The Chinese numerals zero and one to nine that are now in use were adopted before the Tang Dynasty, and the capital forms of the Chinese numerals were first used by people in the Tang Dynasty on formal occasions, such as in official documents. After establishing the decimal system, the ancient Chinese adopted fractions, decimal fractions, and negative numbers, thus greatly expanding their understanding of numbers.

The ancient Chinese were the first to use the decimal system. It is possible that the Chinese created the decimal system because their language depended on characters (like pictures) instead of an alphabet. Historical records show that the ancient Babylonians used symbols similar to later Roman numerals in counting; the ancient Egyptians and ancient Greeks also used special signs for numbers such as 20, 30, 40, which are multiples of 10. For instance, in ancient Greece people used twenty-seven letters to record numbers, with nine letters for numerals from 1 to 9, another nine for 10 to 90, and the other nine for

Asking the Heavens, by Xiao Yuncong, Ming Dynasty. It includes the eight trigrams.

100 to 900. Such clumsy special signs of place values were used in Europe until the eve of the Renaissance. In India they began using the decimal system in the sixth century, while in Europe, the earliest instance of the use of decimal system was probably in a Spanish manuscript dated around 976. The decimal system was, therefore, a great, enduring contribution to mankind from the ancient Chinese. Dr. Joseph Needham once said that without the decimal system it would be most unlikely that the modern world would have developed as it did.

The binary system forms the mathematical basis for computer technology. In its modern form, it was invented by G. W. Leibniz (1646–1716), a multi-talented German philosopher and mathematician, and also an avid Sinophile. Leibniz's binary arithmetic was closely related to the *bagua*, or eight basic trigrams, contained in the Chinese classic *Book of Changes*. The basic trigrams, said to have been invented by one of the earliest legendary Chinese rulers Fuxi, developed by King Wenwang of Zhou, and interpreted by Jiang Ziya (c. 1155–1045 BC), are formed by solid lines (representing *yang*) and broken lines (representing *yin*), and combinations of the eight trigrams constitute the sixty-four hexagrams. Taking the solid line as 1, and the broken line as 0, the sixty-four hexagrams become natural numbers in the binary system in a descending order: 111111 (63), 111110 (62), 111101 (61)… 000000 (0).

When he reached a dead end in creating a multiplier, Leibniz was inspired by the *Segregation-Table* and *Square and Circular Arrangement* supplied by his friend Joachim Bouvet (1656–1730), a French Jesuit scientist and missionary to China. He then went on to refine and develop the first binary system. In 1703 he published *Explication de l'arithmétique binaire, avec des Remarques sur son utilité, et sur ce qu'elle donne le sens des annciennes figures Chinoises de Fohy*, in *Mémoires de l'Académie Royale des Sciences* (*An explanation of binary arithmetic, with remarks about its use and the meaning it gives to the ancient Chinese figures of Fuxi*). Of binary numeration, he writes: "it permits new discoveries in… arithmetic…in geometry, because

when the numbers are reduced to the simplest principles, like 0 and 1, a wonderful order appears everywhere…"

In China, philosopher Shao Yong (1011–1077) of the Northern Song Dynasty put forward a fairly complete binary system in his essays on *Book of Changes*, but his discovery was not disseminated.

Computers, now dominant in so many aspects of modern life, are based on the binary system.

Zu Chongzhi and Pi

Ancient China then was a land of mathematics, and the first to work out the decimal system and the binary system through the yin and yang and "Eight Trigrams" used for divination. These two inventions have had a great impact on contemporary life as well as on the progress of history. The comments by Zhao Shuang on *Zhou Bi Suan Jing* (*The Arithmetical Classic of the Gnomon and the Circular Paths of Heaven*) in the Western Han Dynasty or earlier elaborated the theorem that the sum of the squares of the lengths of the sides of a right triangle is equal to the square of the length of the hypotenuse—the Pythagorean Theorem. The Chinese discovery was made at about the same time as the Greek one, but they were independent, and not related. Liu Hui, a great Chinese mathematician of the Three Kingdoms Period, put forward his own principle about the calculation of the volume of polyhedrons in his *Comments on the Nine Chapters on the Mathematical Art*. Liu also developed a formula for calculating the area of a circle through infinite segments and the idea of the limit, and also a scientific procedure for computing approximate values of the ratio of the circumference of a circle to the diameter. He worked out the area of a regular polygon of 192 sides, and an approximate value of Pi, at 3.14, or 157/50 and 3927/1250 in terms of fractions.

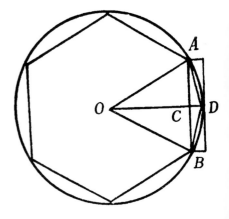

Liu Hui's method of inscribing the circle.

Based on the achievements of Liu Hui and other mathematicians, another great Chinese mathematician made an important breakthrough in the calculation of the ratio of the circumference of a circle to its diameter.

Zu Chongzhi.

Zu Chongzhi (429–500) was born into the family of a scholar-bureaucrat of the Southern Dynasties Period. From his childhood he was diligent, eager to learn, and keen to explore. At the age of twenty-five he entered the Hualin Academy to continue his academic career. Later he worked under a high-ranking official as a staff member, which gave him sufficient time for scientific research. Zu made great progress in the studies of astronomy, calendrical science, mathematics, and mechanics. The achievements of Zu Chongzhi and his son Zu Geng were recorded in their work *Zhui Shu* (*Method of Interpolation*), and this great work was listed among the Ten Mathematical Classics and used by Chinese students of the Tang Dynasty as well as by Korean and Japanese students. It is a great pity that this book has been lost, and Zu's achievements can now be found only in fragments recorded in other mathematical works. These achievements fall mainly into three areas: computing the ratio of the circumference of a circle to its diameter; calculating the volume of globes; and solving higher degree equations.

In calculating approximate values of the ratio of the circumference of a circle to the diameter, the ancient Greeks outdid the ancient Chinese for quite a long time. In the fifth century BC, when Greek mathematicians put the value of the ratio at 3.1416, the Chinese still believed the ratio of the circumference to the diameter was three to one, and used this value until the Han Dynasty. In the Western Han Dynasty, the mathematician Liu Xin calculated two approximate values of the ratio—3.141547 or 3.14166, with the significant figure being 3.1.

Two pages from *Jiuzhang Suanshu Zhu* (*Comments on the Nine Chapters on the Mathematical Art*) by Liu Hui.

In the Eastern Han Dynasty, the scientist Zhang Heng worked out two expressions for approximate values of the ratio: 92/29, and the square root of 10. Liu Hui worked out the value as 3.14, but Zu Chongzhi was not satisfied with this result. Using Liu Hui's method of inscribing the circle, Zu calculated the areas of a hexagon and a 6×2^{12}-gon, and calculated the ratio to be between 3.1415926 and 3.1415927. To obtain such a result, Zu had undertaken extremely lengthy calculations involving hundreds of square roots, all to an accuracy of nine decimal places. This was a work that took great perseverance, resolve, and energy. It was a great achievement of the time, and it was more than 100 times more accurate than Liu Hui's value. Zu devised a precise method to provide a range for the changes of the ratio, the basic method for the expression of an irrational number, a method also used by the great Greek mathematician Archimedes. Zu found two values for the ratio: the "approximate value" of 22/7, and the "so-called precise value" of 355/113 (a better approximation). The precise value is so close that some mathematicians have proved that if it is used to calculate the area of a circle with a radius of 10 km, the error will be within millimeters. In Europe, this "precise value" was obtained by Valentin Otto (German) and Marcus Antonius (Dutch) for the first time in 1573 and was named "Antonius Ratio"—more than 1,000 years later than Zu. Since 1913 it has also been known as the "Zu Ratio."

Equal Temperament

The ancient Chinese led the world in the studies of music and pitch; the earliest record of such studies was found in *Guanzi*, a book written during the Warring States Period. The studies found that of the five tones—*gong, shang, jue, zhi, yu*—on the ancient Chinese five-tone scale, the sound-wave frequencies of the upper and lower notes formed simple mathematical ratios of 3:2 or 4:3, known as the three-section. A scale comprises a number of fifths, known as the circle-of-fifths. Adding two semitones to the five tones makes a seven-tone scale. The circle-of-fifths system led to the development of a 12-tone system, which corresponds to c, c#, d, d#, e, f, f#, g, g#, a, a#, b, c1 in Western musical terms.

According to the circle-of-fifths principle, after eleven circles the sound-wave frequency of the last tone should be twice that of the first tone. But in reality that is not the case, which was a great puzzle to the ancient scholars who studied music and pitch, and spurred them on to further exploration. Ancient Chinese researchers had devoted their efforts to finding the absolute values of the twelve tones and the accurate ratios of upper and lower tones. Chimes of bells from the Spring and Autumn Period were unearthed in 1957 in Xinyang, Henan Province and also excavated in Suixian County, Hubei Province in 1978. The pitch and sound-wave frequency of each bell are close to those

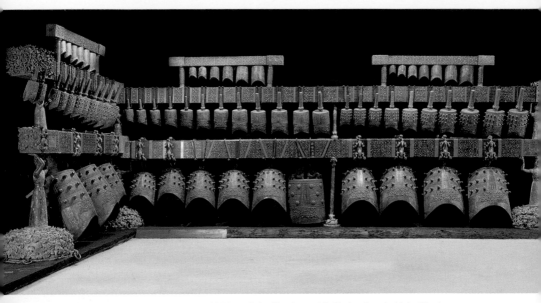

Set of bronze bells unearthed in 1978 from Duke Zeng's tomb in Suxian County, Hubei Province. The bells produce a wide range of fine tones.

determined by present-day twelve-tone equal temperament. But they were only improvements from the three-section method, and not entirely accurate. From the Han, Wei, Jin, Sui, and Tang Dynasties to the Five Dynasties and the Song Dynasty, the search for accurate sound-wave frequencies continued, but the puzzle was not solved. Then a genius, Zhu Zaiyu, finally appeared, and solved the problem.

Zhu Zaiyu (1536–1611), a noted musician, mathematician, and astronomer-calendarist of the Ming Dynasty, was born in Henei County of Huaiqing *Fu* (now Qinyang City, Henan Province). He was a Prince of Zheng, and the ninth-generation descendant of Zhu Yuanzhang, the first Ming emperor. His father, however, was persecuted and sent to prison, and the family fell into poverty. Despite this bad fortune, he devoted all of his energy to science, and especially to the study of music and pitch. When the Emperor decided to exonerate him and resume his royal family status, Zhu refused, and continued his studies. By instinct and training, Zhu developed a musical talent. In 1560 he finished his first work, *Scores for the Se* (lute). In 1584 he completed the masterpiece, *Lüxue Xinshuo* (*New Explanation of the Theory of Pitch*), in which he proposed the theory of equal temperament and a method of calculation. This was a brilliant achievement in both the Chinese and world history of music. Zhu's discovery came more than one hundred years earlier than his counterparts in the West.

Zhu Zaiyu not only worked out the theory of equal temperament, but also put it into practice and created the world's first stringed instrument based on this theory—the *xian zhun*, which produces twelve accurate tones. Applying the extraction of square and cube roots to the study of musical tuning, Zhu discovered the difference in sound-wave frequencies of two neighboring tones to be the number 2 extracted for 12 times with an accuracy of more than 10 decimal places. The equal intervals ensure ideal musical modulation, and this revolutionary discovery has proved to be scientifically accurate. Zhu said: "Doing away with the three-section, the new method enables

Two good friends, Yu Boya and Zhong Ziqi, meeting to enjoy *qin* music. Yuan Dynasty scroll painted by Wang Zhenpeng.

unending circulation of twelve tones in an orderly way, and this is an unprecedented achievement in the two thousand years of the history of music." Musicians in other countries have tested Zhu's method, and his theory of equal temperament is now used throughout the world.

Apart from musical tuning, Zhu Zaiyu's works also cover mathematics and astronomy, and he achieved a great deal in all three disciplines.

Other Major Inventions (III)

我国明代醫藥學家李時珍善於採集和整理各
種藥物，曾經二十八年以上的時間，走遍大江南北，
踏訪名山大川，嘗百草，訪民間，作《本草綱目》，
乃我國藥學三聖經典鉅著之探源尋趣精究寶典
无實九年九月李武養作于南京

Li Shizhen collecting medicinal herbs.

Traditional Chinese Medicine

Before Western medicine came to China, traditional Chinese medicine, a system unique to China, had been the major guarantee for people's health in the country for several thousand years. Nowadays it is still a major means of healthcare for the Chinese, and its advantage is shown in the case of some serious conditions that modern medicine fails to treat. Traditional Chinese medicine has its own concepts, means of diagnosis, treatment, and composition of drugs and prescriptions, and these are quite different from those of modern Western medicine. Its basic theory and practices were established as early as in the Spring and Autumn and Warring States Periods. Its philosophical basis is the principle of integration of man and nature. It regards the human body as a whole, and man and nature as a whole, and believes that any disorder of the human body reflects discord between man and nature, or between the patient and the outside world.

In the theory of traditional Chinese medicine, the human body contains five organs of *zang*, and six entrails of *fu*. Each of the *zang* and *fu* has its own functions, and also controls a certain aspect of the normal operations of the human body. The *jing luo* (meridian) system connects various parts of the human body, and allows *qi* (a vital substance that flows throughout the body), blood, and fluids to circulate. Unlike the nervous system in Western medicine, the *jing luo* system is invisible, yet the mysterious channels and collaterals function quite effectively in transmitting vital substances. In traditional Chinese

Five zangs and six fus
The five organs of *zang* refer to the heart, liver, spleen, lungs, and kidneys, while the six entrails of *fu* include the gall bladder, stomach, small intestine, large intestine, three cavities (*san jiao*), and bladder. The three cavities refer to the part of the body cavity above the diaphragm (upper *jiao*), the upper part of the abdominal cavity (middle *jiao*), and the lower part of the abdominal cavity (lower *jiao*). In general, *zang* refers to the substantial organs in the thoracic and abdominal cavities with the common function of storing *jing* (essence) and *qi* (vital force). The *fu* refers to the organs with cavities in the thoracic and abdominal cavities with the functions of digesting food, absorbing nutrition, and voiding excreta. Besides being used as labels in anatomy, more importantly, *zangs* and *fus* in traditional Chinese medicine can also be used to summarize the physical functions and pathological changes of the human body. Though having the same names as those in modern medical science, the *zang* and *fu* organs of traditional Chinese medicine are totally different in concept and functions, and are thus not fully equivalent.

medicine, the various systems of the human body are closely and complexly related in a comprehensive life system, which functions as a whole. This then is the basis of the physiology and pathology of traditional Chinese medicine.

In diagnosis, traditional Chinese medicine relies on four methods: listening and smelling; enquiring; observing; and pulse-feeling. Listening and smelling involves listening to the patient's voice, breathing and coughing, and smelling the odor of the body and of excretion products. Enquiring involves asking about the

Instruments used for processing medicinal herbs.

patient's case history. Observing involves assessing the patient's mental state, facial expression, complexion, and the color of the tongue, fingers, and nails. In pulse-feeling, the doctor puts his index finger, middle finger, and third finger on the wrist of the patient and feels the pulse to ascertain the symptoms and causes of the disorder according to certain conditions of pulse: frequency, rhythm, fullness, evenness, and amplitude. Through the years Chinese doctors accumulated rich experience in pulse feeling. According to legend, an experienced doctor could feel the pulse of a patient through a string. In feudal times, women of royal or noble families could not meet strange men face to face, let alone to be touched by a man. Therefore when such a woman fell ill, a string would be tied to her wrist and a doctor hidden behind a screen would "feel" her pulse at the other end of

Legend has it that an experienced doctor could feel a female patient's pulse through a string.

the string. The theory of *yin* and *yang* and the five elements, a complicated system of knowledge and methods, guides the diagnostic practice of traditional Chinese medicine. For instance, the symptoms of fever, excitement, quickened pulse, reddish skin, and thirst belong to yang, while the symptoms of coldness in the hands and feet, pale skin, slowed pulse, and weakness belong to *yin*. *Yin* and *yang* are interdependent and can also transform to the opposite. A patient's conditions are analyzed and differentiated in accordance with the eight principal syndromes: *yin* and *yang*, deficient and excessive, exterior and interior, and cold and heat. With the symptoms decided, medications

Over thousands of years, Chinese doctors have accumulated many proven recipes, now used not only in China, but also in many other parts of the world.

are correspondingly prescribed to achieve best treatment. The five elements are wood, fire, metal, water, and earth. These are inter-promoting: wood promotes fire; fire promotes earth; earth promotes metal; metal promotes water; and water in turn promotes wood. The five organs of *zang* and six viscera of *fu* are also classified by the five elements, and they are believed to influence each other.

The Chinese *materia medica* classifies the properties of drugs into four groups: cold, hot, warm and cool; and drugs of five tastes (sour, sweet, bitter, pungent, and salty) are used to treat different diseases. The Chinese *materia medica* is derived from nature. *Shennong Ben Cao Jing* (*Shennong's Herbal Classic*), the earliest extant work on *materia medica* in China, listed 365 kinds of drug, including 252 from plants, 67 from animals, and

46 from minerals. The hardest and the most mysterious parts of traditional Chinese medicine are diagnosis and treatment. These are based on an overall analysis of symptoms and signs to establish the cause, nature, and location of the illness and the patient's physical conditions according to the basic theories. In prescribing drugs, combinations of drugs are different from patient to patient, from illness to illness, and even between phases of the illness. Chinese doctors have accumulated numerous proven recipes, based on thousands of years of experience, and these are treasures not only for China but also for the world at large. In making prescriptions, the principle of combining the "monarch, minister, assistant, and guide drugs" is followed. The monarch drug is the principal part of the recipe, which produces the main effect to treat the major cause and symptoms of a disease; the minister drug is secondary, and helps the principal drug; the assistant drug serves to reduce the unwanted side effects of the principal and secondary drugs; and

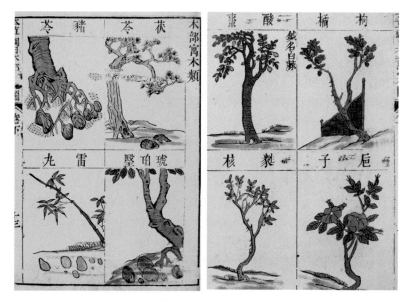

Two pages from *Bencao Gangmu* (*Compendium of Materia Medica*).

the guide drug directs the action of a prescription to the affected parts of the human body, to reinforce the treatment. Also taken into consideration in selecting the ingredients of prescriptions are the seven different effects in compatibility, namely: using alone, mutual reinforcing, assisting, incompatibility, inhibition, detoxifying, and antagonism.

Ancient China featured many great men of medical science. The most famous doctors were Bian Que (407–310 BC) of the Spring and Autumn and Warring States Periods, Hua Tuo (?–208) and Zhang Zhongjing (c.150–219) of the Eastern Han Dynasty, and Sun Simiao (581–682) of the Tang Dynasty. *Shang Han Za Bing Lun* (*The Treatise on Febrile and Miscellaneous Diseases*) by Zhang Zhongjing, and *Qian Jin Yao Fang* (*Essential Prescriptions Worth a Thousand in Gold*) by Sun Simiao are two essential works for practitioners of traditional Chinese medicine. The most famous practitioner of Chinese *materia medica* is Li Shizhen (1518–1593) of the Ming Dynasty, whose *Bencao Gangmu* (*Compendium of Materia Medica*) includes 1,892 medicinal substances, 11,000 prescriptions, and 1,100 illustrations. This authoritative work has been translated into various languages, and distributed in many countries.

Acupuncture

Acupuncture, an ancient Chinese medical procedure now spread to many parts of the world, is still rather mysterious. The *jing luo* or meridian system, on which acupuncture is based, has never been verified by any anatomic substance nor demonstrated by any medical technique. The numerous oddly named acupuncture points are in fact virtual dots on the skin. The practice, however, can quickly bring about positive results, and in some cases even instant cures. It is now impossible to discover who invented this miraculous medical technique, or when it was invented.

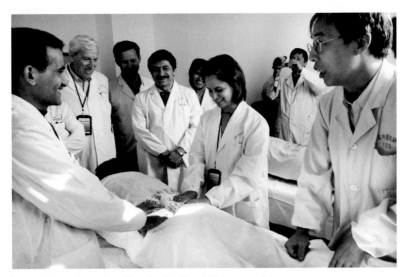

Visiting doctors from abroad learning traditional Chinese medicine.

Historical records of acupuncture link it to the earliest legendary rulers of Huangdi and Fuxi. *Zhen jiu*, the Chinese for acupuncture, actually means both acupuncture (*zhen*) and moxibustion (*jiu*). As acupuncture is far more effective than moxibustion, nowadays people tend to regard *zhen jiu* simply as acupuncture.

Acupuncture was probably invented by accident. Ancient people might have accidentally found that they could relieve pain by pricking certain points of the body with a sharp stone. In Chinese classics such as *Shan Hai Jing* (*The Classic of Mountains and Seas*), *Shuo Wen Jie Zi* (*The Original of Chinese Characters*), and *Zuo Zhuan* (*Zuo's Annals*), there are records of using a sharp stone to prick the body to relieve pain. Later on, stone needles were made for acupuncture and eventually needles of other materials were developed: bone, bamboo, wood, bronze, iron, gold, and silver. The earliest gold and silver needles were unearthed from a Western Han tomb in Mancheng County, Hebei Province.

Song Dynasty painting by Li Tang, showing acupuncture.

The technique of acupuncture developed rapidly in the Spring and Autumn and Warring States Periods. *Huangdi Nei Jing* (*Yellow Emperor's Classic of Internal Medicine*) contains systematic discussions about acupuncture. Zhang Zhongjing and Hua Tuo, two noted doctors of the Han Dynasty, were also experts in acupuncture, and their practices are found in historical records. In the meantime, there were also excellent acupuncture practitioners among the common people. *Hou Han Shu* (*The History of*

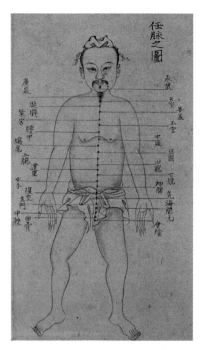

Acupuncture points.

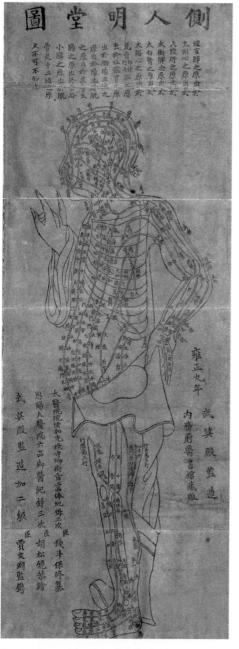

側人明堂圖

Acupuncture points mapped on the body.

the Later Han Dynasty) tells about an old fisherman in Sichuan who fished on the Fujiang River and volunteered to treat local residents with his outstanding technique of acupuncture. This proves that at that time acupuncture was quite popular, and also widely practiced.

To understand the theory and practice of acupuncture, it is necessary to know something about *jing luo*. The *jing luo* (meridian) system consists of *jing* (channels) and *luo* (collaterals) and other subsystems, closely related to the five organs of *zang* and six viscera of *fu*. The channels are the main conduits for the circulation of *qi* and blood, while the collaterals are a superficial network connecting the channels as well as all portions of the body. On the channels and collaterals are numerous acupuncture points. By applying a needle or ignited moxa to the points, special sensations will be transmitted along the channels and collaterals, and in this way illness or disorder related to the channels and collaterals is treated. Although the induced sensations are real, modern anatomy has never found any trace of the channels, collaterals, or acu-points.

Huangfu Mi (215–282), a celebrated physician of the Jin Dynasty, contributed a great deal to the development of acupuncture. Huangfu once tried to make pills to confer immortality and become seriously ill. After his recovery, he devoted all his efforts to medicine, especially the study of acupuncture. He compiled the 32-volume, 128-chapter *Zhen Jiu Jia Yi Jing* (*The Systematic Classic of Acupuncture and Moxibustion*), which later became an essential reference for the study of acupuncture. This work discusses human physiology, the channels and collaterals, acupuncture points and their functions, methods of applying the needles, and pathology of the human body. It also discusses in detail the clinical treatment of various diseases, gives unified definitions of the channels, collaterals and points, and explains the dos and don'ts of acupuncture. The classic was introduced to other countries quite early and in the eighth century the Japanese began to use it as a medical textbook. There is now an English edition of the work, and a French version is in preparation.

In the early years of the Northern Song Dynasty, Wang Weiyi (987–1067), a master practitioner of acupuncture, compiled the three-volume *Tong Ren Shu Xue Zhen Jiu Tu Jing* (*Illustrated Manual of Acu-points of the Bronze Figure*), and created two bronze figures. The life-size male figures can be taken apart to show the organs and viscera, and 14 channels and collaterals, and 657 acu-points are marked on the body. According to legend, the bronze figures

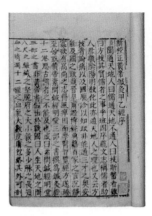

Page from *Zhen Jiu Jia Yi Jing* (*A-B Classic of Acupuncture and Moxibustion*), the earliest surviving work on acupuncture, by Huangfu Mi, a noted Jin Dynasty physician.

Ming Dynasty bronze figure used for teaching acupuncture. It is 213cm tall and has 666 acupuncture points.

were used to teach and test acupuncture students. When a test was conducted, the figure was filled with mercury and the acu-points on the body were covered by wax. When the student made a correct puncture, the mercury would flow out of the acu-point. Later on more bronze figures were made, and some were sent abroad. Now one bronze figure is kept in a museum in St. Petersburg in Russia and another is in the Imperial Palace in Japan.

Anesthetic

Anesthetics are necessary for surgical operations, and it is difficult to imagine an operation without one. However, in the early nineteenth century there were no reliable anesthetics in Europe, and Napoleon's surgeon was said to have had to rely on speed to minimize the pain of wounded soldiers, and was said to have performed amputations on more than 100 wounded soldiers in a single night.

In China, as early as the Warring States Period, Bian Que, a famous physician and surgeon of the time, had concocted an

anesthetic "toxic wine" to be used in surgical operations, and in the third century, Hua Tuo (?–208) invented a general anesthetic for a patient to have an abdominal operation. Hua Tuo was a celebrated physician and surgeon of the Eastern Han Dynasty, and practiced medicine in the provinces of Anhui, Jiangsu, Shandong, and Henan. When drugs and acupuncture failed to cure illness in the internal organs, Hua Tuo would perform operations on patients. Before the operation, he would ask the patient to take the anesthetic with wine. When the patient

Hua Tuo.

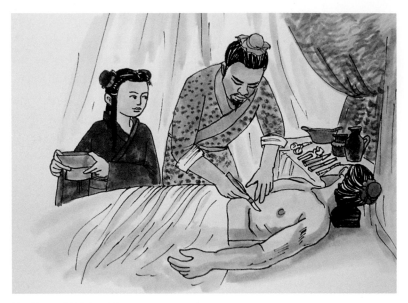

Hua Tuo performing surgery.

lost consciousness, Hua Tuo would open the chest or abdomen to treat the affected organ and then apply a magic ointment, and the patient would recover within four or five days.

Legend has it that before his invention of anesthetic Hua Tuo had performed an unconventional operation on Guan Yu (?–220), a heroic general of the Shu state. During battle, General Guan was shot in the arm and the arrow that hit him proved to be poisoned. The poison quickly spread to the bone, and Guan's men sent for Hua Tuo. After examining the wound, Hua told Guan that it was too late for any drug or acupuncture to work, and the only way to save him was to cut open the wound and remove the poison from the bone. Without any hesitation Guan stretched out his arm for Hua, while continuing a game of chess with another person. Guan remained calm, talking and laughing, while Hua used a knife to scratch the poison from the bone. Knowing that ordinary people would never be able to endure such great pain, Hua worked even harder, and finally perfected his invention.

When Cao Cao (155–220), a great strategist who ruled northern China at that time, had a severe headache he sent for Hua Tuo. Hua examined Cao and told him that he should take anesthetic first, after which he operated and removed the cause of the headache from his head. Cao suspected that Hua intended to kill him by applying the anesthetic, and executed Hua in cold blood. Hua's anesthetic was thus lost.

According to the *History of the Later Han Dynasty*, Hua used his anesthetic in surgical operations such as removing part of the intestines, caesarean births, and excising tumors. Even today such operations are not minor ones. Hua's anesthetic technique had a great influence on later medical practitioners. In the Song, Yuan, Ming, and Qing Dynasties, Chinese doctors developed agents for local anesthesia, but they failed to discover the formula for Hua's anesthetic. The only known ingredient was stramonium, which causes loss of sensation. Stramonium is derived from

Jimsonweed or Thornapple (*Datura stramonium*), used in ancient China as an anesthetic.

Jimsonweed or Thornapple (*Datura stramonium*), a poisonous plant in the potato family, *Solanaceae*.

More than a century ago, a Japanese doctor claimed that he had found the formula of Hua's anesthetic, which included stramonium, chuanxiong rhizome, aconite, and arisaema tuber. However, when the formula was tested on two people, one was killed and the other blinded. It seems that Hua's formula may remain a secret.

In the West, American dentist Horace Wells was the first to use laughing gas (nitrous oxide) as an anesthetic in 1844, but the result was not satisfactory. In 1848 William Morton, another American dentist, publicly demonstrated the use of ethyl ether as a general anesthetic during surgery.

Variolation

Smallpox, an acute and highly contagious disease, has inflicted much damage on humanity and the procedure known as variolation (an early form of vaccination) marked a breakthrough in the fight against this epidemic virus. This great achievement in immunology was introduced by the Chinese no later than the sixteenth century. Variolation and vaccination finally led to Jenner's invention in 1796, and the eradication of this fatal disease as declared by the World Health Organization in 1979.

Variolation was developed from the Chinese tradition of disease prevention. In the *Book of Changes*, the *Yellow Emperor's Canon of Internal Medicine*, and other classics, a number of thoughts were expressed about attitudes to disease prevention: "The wise man should pay attention to the prevention of diseases;"

Sun Simiao.

"An excellent doctor usually treats people before they apparently fall ill, therefore illness rarely occurs;" "Is not it too late to administer drugs when illness has developed?;" "By detecting delicate changes, a good doctor is able to treat a patient as early as possible to cure the disease and help the patient recover." Chinese doctors also developed the concept of "combating poison with poison." Ge Hong (c. 281–341), a well-known doctor in the third century, used to cut a piece of the brain of a deceased mad dog to cover the wound of the patient who was bitten by the dog, with the intention of preventing rabies. Sun Simiao (581–682), a celebrated medical scientist of the seventh century, used a small knife to transfer the blood and pus of a patient with an ulcer to healthy people, placing this under the skin as a means of prevention. These were just some of the experiments and practices in immunology in ancient China.

Smallpox was said to have spread to China from the south in the second century BC, and Chinese doctors made great efforts to fight and prevent this dreadful disease. When did the Chinese invent variolation? There seem to be three possible answers to this question: a person with the surname Zhao, in the Tang Dynasty invented it in the eighth century; doctors at Mount Emei in Sichuan, in the Song Dynasty developed the method in the eleventh century; and people in Anguo, Anhui Province, in the Ming Dynasty practiced variolation in the sixteenth century. Scholars tend to accept the third version, but they also acknowledge that the first two are not necessarily groundless. In all three cases, the principle involved combating poison with poison: infecting people with substance from the pustules of smallpox patients boosted immunity to the fatal disease.

Emperor Kangxi of the Qing Dynasty contributed a great deal to the popularization of variolation, and

Girl suffering from smallpox.

in 1681, he sent envoys to Jiangxi Province in the south to recruit doctors practicing this technique. Zhu Chungu was the first doctor sent by the Emperor to inoculate the children of Manchu and Mongol officials in the northeast and Khalka, a program that met with great success. Emperor Kangxi later commented with joy: "In the beginning of this dynasty, people were very afraid of smallpox. Since I got the method of variolation, your children have been protected from the disease. I ordered the practice to be carried out in the forty-nine banners in the northeast and all the feuds in Khalka. All people inoculated are well protected. I still remember that when this was first done, many elderly people were quite surprised. I have insisted on it, and thousands upon thousands of people have been saved." Zhang Yan, a contemporary of Zhu Chungu, reported: "I have variolated more than 9,000 people, and the failures numbered only 20 to 30." The rate of success was thus 97 to 98 percent.

Through Emperor Kangxi's efforts, the spread of smallpox was curbed in China. When the good news reached other countries, Russia sent doctors to China to learn the technique in 1688. The method was popularized in England in 1721–1722 by Lady Mary Wortley Montagu, and won the support of the Queen.

In the seventeenth century, the technique of variolation spread from China to the Americas, and also to Japan and Korea where it was popularized in the early eighteenth century. In 1796 the British physician Edward Jenner (1749–1823) discovered vaccination, and used the cowpox virus to confer protection against smallpox, a related virus, in humans. Vaccination was introduced to China in 1805. Both vaccination and variolation have contributed to the elimination of smallpox around the world.

The Great Wall

The Great Wall is a symbol of China—a testament to the long history of the country and a crystallization of the wisdom, strength, and will of the ancient Chinese. In 1987, the Great Wall was listed as a World Heritage Site by the United Nations Educational, Scientific and Cultural Organization (UNESCO).

In the Warring States Period, various states built their own walls to guard against invaders, and these walls were located in the valleys of the Yellow and Yangtze rivers. After unifying China, Emperor Shihuang of the Qin Dynasty dismantled most of these walls, and shifted the defense to the north, against the Huns. The Emperor sent General Meng Tian (?–210 BC) on a northern expedition, and began construction of the Great Wall. Making use of the geography, the general built watchtowers and fortifications along the Yellow River and the Yinshan Mountains, and connected the existing walls of the former Yan and Zhao States in the north and the east and those built by King Zhaowang of the smaller Qin State in the west. The completed Great Wall extended from the Liaodong Peninsula in the east to Lintao of Gansu Province in the west, and measured more than 10,000 *li* (one *li* is about 0.5 km) in length. In the Han Dynasty,

Ruined section of the Great Wall dating from the Han Dynasty, near Dunhuang, Gansu Province, northwest China.

The Great Wall—widely seen as a symbol of China.

Emperor Wudi, who ruled from 133 to 87 BC, ordered generals Wei Qing and Huo Qubing to attack the Huns, repair the Great Wall and add new sections to it, thus extending the Great Wall more than 2,000 *li* westward, to the west of Jiuquan and Dunhuang, in Gansu Province.

To protect the country against Tartars from the north and Nuchens from the northeast, the Ming rulers continued building the Great Wall. For more than 200 years, the Ming Empire implemented eighteen large-scale construction projects and built a Great Wall of 6,700 km, extending from Jieshi in the east to Jiayu Pass in the west.

As a comprehensive defense system, the Great Wall consists of passes, walls, watchtowers, and beacon towers. The wall itself was usually built along mountain ridges. The western parts of the Great Wall are mostly of rammed earth, while the sections near Beijing are made of stones and bricks. Badaling, the section of Great Wall north of Juyong Pass, lying more than 1,000m above sea level, was built with stone blocks and large bricks. Averaging 7.8m in height, the wall measures 6.5m wide at the base and 5.8m at the top, wide enough for five horses or ten people to march in a row. On top of the wall there are parapets, buttresses, holes for observation and shooting, and there are watchtowers and beacon towers at intervals of several hundred meters.

Each watchtower can hold several dozen soldiers, and the beacon towers were used for lighting signal fires to transmit information. There are about 200 passes along the Great Wall, from Jiayu Pass in the west to Shanhai Pass in the east. Other famous passes include Juyong, Yanmen, and Zijing. In a section near Shanhai Pass, the Great Wall crosses a river in a valley, and here there are nine arches in the wall to allow the flow of water. In modern times, the passes no longer serve as defense fortifications. Some of them have become thoroughfares, some

The Great Wall at Badaling, near Beijing, constructed with stone blocks and large brinks.

have developed into cities, and others have become historic sites and tourist attractions.

The Great Wall was the most time-consuming and labor-intensive construction project in human history, and is estimated to have taken 180 million cubic meters of rammed earth and 60 million cubic meters of stones and bricks—more than enough to build a wall five meters high and one meter wide around the globe. To build such a gigantic structure in the mountains, it is difficult to imagine how the sites were surveyed, how the wall and other facilities were designed and built, where the materials were obtained, how they were transported to the sites, and how the construction workers were supplied.

China's Grand Canal

China's Grand Canal runs from Beijing in the north, to Hangzhou in the south and is the oldest and longest canal in the world. As the greatest engineering project in water resources and transportation in ancient China, this waterway is nearly as famous as the Great Wall. The earliest section of the canal is the Han Canal, built in 486 BC. In the later years of the Spring and Autumn Period, King Fucha of the State of Wu fought his rivals in the north, and set up the city of Han (today's Yangzhou). Later he built the Han Gou (Han Canal) to divert water from the Yangtze River

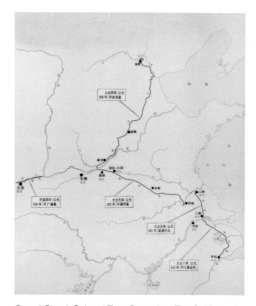

Grand Canal, Sui and Tang Dynasties. The Sui Dynasty canal, running 2700 km from Zhuojun in the north to Yuhang in the south, played a vital role in economic and cultural exchange.

in the south. A number of man-made ditches connected lakes along its way, and the Yangtze, Huaihe, Yishui, and Jishui Rivers. In the Sui Dynasty, Emperor Yangdi dispatched contingents of builders to work for six years on a 2,700-km Grand Canal, which ran to Zhuozhou in the north and Hangzhou in the south, from the Sui capital Luoyang. The ancient Han Canal was dredged and widened to become an important section of the Grand Canal. In the thirteenth century, Kublai Khan, the first emperor of the Yuan Dynasty, set up the Yuan capital in Beijing. To connect the economic center in the south with the political center in the north, from 1283 to 1292 the Yuan Empire implemented a gigantic project to revamp and improve the Sui Dynasty Grand Canal. The new Grand Canal, with a total length of 1,792km, ran from Tongzhou in the suburbs of Beijing to Hangzhou, now capital of east China's Zhejiang Province.

The Grand Canal connects the six municipalities and provinces of Beijing, Tianjin, Hebei, Shandong, Jiangsu and Zhejiang, and the five water systems of the Haihe, Yellow, Huaihe, Yangtze, and Qiantang rivers. The Grand Canal consists of man-made waterways, rivers, and lakes, and can be divided into seven sections. As an artery between the north and the south, the Grand Canal has played a vital role in Chinese history. The waterway was the lifeline for the feudal regimes, which depended on it to ship grain supplies from the south to the north. Besides grain, the canal also transported other commodities. The regions along the canal therefore developed into an extensive economic belt, as boats brought businessmen and travelers to the towns and cities on its banks. In the Ming and Qing Dynasties, the markets along the canal were quite prosperous.

Such a grand engineering project required fairly sophisticated technology. First, large-scale topographic surveys had to be carried out along the line, and the canal also had to cross some mountains and hills. Because of the uneven topography and the needs of navigation, locks had to be built to control the water

Detail from *Traffic on the Grand Canal*, a Qing Dynasty scroll painting showing busy traffic on the Tongzhou section.

level; the Grand Canal is noted for its numerous locks, and the construction workers accumulated rich experience in building these. Although there is no record about when the first lock was built, a book compiled in the Northern and Southern Dynasties records that during the Jingming reign of the Southern Song Dynasty (423–424) someone was drowned in a *dou men* (lock) when sailing through Yangzhou, thus indicating the existence of locks at that time. In the Tang Dynasty, eighteen locks were built on the Lingqu, another canal, and in the Northern Song Dynasty the technology of building locks was further developed. At that time, Qiao Weiyue (926–1001), a deputy governor in charge of transportation in Huainan, built several locks on the Grand Canal to facilitate the shipping of grain supplies from the south. In 984 he supervised the construction of Xihe Lock, which was seventy-six meters long, with two movable gates at both ends. When the gates were lifted alternately, the water level in the lock would change, to allow the passage of boats. According to a record made in 1118, more than 500km of waterways on the Grand Canal depended on locks to ensure navigation. In Europe, the first lock was built in 1373.

The Gongchen Bridge over the Grand Canal in Hangzhou. The high, stone-arched bridge greeted emperors arriving on their tours of inspection.

The Chinese people greatly value the Grand Canal, and China has applied for its inclusion in UNESCO's World Heritage List.

Dujiangyan Irrigation System

China encompasses a large range of climates, and also has a complex topography. Over much of the country the climate is very changeable, with frequent droughts and floods. Water control projects were built in ancient China to tackle such disasters, and these included canals, ponds, dikes, and underground culverts. The most important irrigation systems among them were canals in the plains and ponds in hilly areas. There were a number of ancient irrigation systems in Shaanxi, Hebei, Sichuan, Beijing, and Ningxia. The most famous among these is the Dujiangyan irrigation system in Sichuan, which still plays an important role in local agriculture.

Dujiangyan, or Dujiang Weir, previously known as Jin Di (Golden Weir) or the Great Weir of Du'an, was built in the third century BC, receiving the name Dujiangyan in the Song Dynasty. It is located in Guanxian County on the middle reaches of the

Minjiang River. Flowing from high mountains and valleys, the river enters the Chengdu Plain, where it slows, depositing large amounts of sand and silt on the riverbed. Flooding was quite frequent. In his later years, King Zhaowang (306–251 BC) of the Qin State appointed Li Bing, a noted water resources expert, to be governor of Shu (Sichuan). Immediately after assuming this post, Li Bing began presiding over the water control project. The irrigation system consists of three parts: Yuzui, Baopingkou, and Feishayan. Li first had the builders cut Baopingkou, an opening in the eastern bank of the Minjiang, intending to divert water to the fields in the Chengdu Plain. However, because of the high terrain, the water was not able to flow through the opening. He then asked the builders to create a structure in the shape of a fish's mouth (*yuzui*) in the middle of the river using bamboo

The Dujiangyan irrigation system, built in the third century BC, provides both irrigation and flood control.

cages filled with stones, thus helping to divert water to the east bank. This structure divides the river into two channels: *waijiang*, the outer main channel, which accommodates floodwater, and *neijiang*, the inner channel, which diverts water for irrigation.

Baopingkou, the opening in the east bank between Yuleishan and Lidui, controls the flow from the inner channel to the irrigation canal. Cutting this opening was a major feat as the rocks there were very hard. The workers first lit fires on the rocks to heat them, and then poured cold water or vinegar on them, causing them to crack. Inch by inch they managed to cut a passage 20 meters wide, 40 meters deep, and 80 meters long. Feishayan, or the spillway weir, was built between Yuzui and Baopingkou. During floods the water flowed over this weir from the inner channel to the outer channel, carrying lots of silt and

"Dredge the riverbed and build a low weir"—the principle for the Dujiangyan project as laid down by Li Bing and engraved on a temple wall in memory of Li Bing and his son.

cobbles and thus reducing the sediment load in the canal. As Dujiangyan is situated at the highest spot in the Chengdu Plain, an alluvial plain that fans out in the Sichuan Basin, the natural flow from the irrigation system reaches large areas of farmland, turning the Chengdu Plain into "a land of abundance."

After more than 2,200 years, Dujiangyan still functions well as a flood control and irrigation project, and has also become a well-known tourist destination. As a representative of ancient engineering in water control, Dujiangyan has been inscribed on the UNESCO World Heritage List.

Miscellaneous Smart Inventions

The inventions and discoveries we have discussed have a bearing on the economy and people's livelihoods, as well as on the development of science and technology in China and in the world at large. Apart from those major inventions and discoveries, there are also other interesting inventions that are still used in daily life, both in China and abroad. These clever inventions have retained their charm in spite of developments in science and technology, and their influences are everlasting. They are products of inspiration, exploration, creativity, and devotion.

Kite

China is the homeland of the kite (also known as "paper hawk"), the oldest heavier-than-air craft that gains lift from the wind. It is believed that the kite was invented some 2,000 years ago by Lu Ban (also known as Gongshu Ban, c. 507–c. 444 BC), a Chinese master carpenter of Lu State in the Spring and Autumn Period. It is said that Lu Ban made a flying magpie out of bamboo pieces. The master carpenter was also the first to use the kite in military reconnaissance. A historical record said: "Gongshu Ban made a wooden hawk and used it in reconnaissance of the Song city." In 549 a rebellion force surrounded the palace of Emperor Wudi of the Southern Liang Dynasty (502–557). The maids of honor in the palace "made a paper hawk and flew it to tell imperial troops about the emergency." This was the start of the use of kites to transmit military information.

The Chinese name for kite, *feng zheng* (*feng* means wind and *zheng*, a musical instrument similar to

Swallow kite.

Dragonfly kite.

zither), was attributed to Li Ye, a minister of Emperor Yindi of the Han Dynasty. Minister Li flew a paper hawk on a string inside the palace, and a bamboo flute on the head of the paper hawk produced a sound like the zither when the wind passed through it. The device was thus named *feng zheng*. In the Ming Dynasty, the Ming troops made a crow-shaped kite filled with gunpowder, and used this to bomb enemy camps. At that time the kite was also used to measure the velocity of the wind. British historian of science, Joseph Needham, listed the kite among major Chinese discoveries. In the National Air and Space Museum in Washington D.C., a sign states that the earliest flying craft were the Chinese kite and rocket.

The kite shifted from military purposes to entertainment in the mid-Tang Dynasty and a Tang poet wrote about children "flying paper hawks." In the Song Dynasty, flying kites became very popular. Emperor Huizong not only advocated the pastime, but also presided over the compilation of *A Collection of Kites from the Xuanhe Years*. Kites were introduced to other countries in Asia, the Arab world, and to other parts of the world, as a popular artifact and toy. In Europe, it was in 1589 that Italian architect

Children Flying Kites, a lunar New Year picture printed at Yangliuqing, Tianjin.

Giacomo della Porta first mentioned the kite in a book. Now the annual international kites fair in Weifang, Shandong Province, in eastern China, attracts kite fans from all parts of the world.

Abacus

The abacus was developed using counters or chips. In the Spring and Autumn Period, chips were widely used in calculation; chips of different colors represented different values, and they were placed in lines and columns to perform arithmetic calculations. Counters or chips were used in calculations in China for 2,000 years, but as life progressed they were found not suitable for more complex calculations. The abacus was thus created.

The principles and methods used in abacus calculations or calculations with beads are much the same as those used in calculations with chips, but the operation is faster and more convenient. The term *zhu suan* (calculation with beads) was first found in a work by mathematician Xu Yue (?–220) of the Eastern Han Dynasty in the second century AD. In *Qingming Shang He Tu*, a panoramic painting of life along the Bianhe River at the Qingming Festival by noted painter Zhang Zeduan (1085–1145) of the Northern Song Dynasty, an abacus can be seen on the counter of a pharmacy. Writer Tao Zongyi (1316–?) of the Yuan Dynasty first wrote about "the beads of abacus," saying, "they are moved by hand." In *Lu Ban Mu Jing*, a book on Ming Dynasty carpentry, there were quite a number of specifications for making abacuses. There were also books on the principles and formulae used in abacus

Abacus
A Chinese abacus consists of a wooden frame, with beads on parallel wires, and a crossbar set perpendicular to the wires, dividing the beads into two groups. Each column—that is, each wire—represents one place in the decimal system. The column farthest to the right is the units column; the next column to the left is the tens column; and so on. In each column, there are five beads below the crossbar, each of which represents one unit, and two beads above the crossbar, each of which represents five units.

Cheng Dawei.

Detail from *Qingming Shanghe Tu*, a panoramic painting showing life along the Bianhe River at the Qingming Festival, by Zhang Zeduan, a famous artist of the Northern Song Dynasty.

calculations. All this suggests that the abacus was widely used in ancient China. The abacus as a calculating device is suitable for almost everyone, even illiterate people.

Historical records show that different abacuses were developed in quite a few countries, but the Chinese abacus inherited the features of calculation with chips: the nine numbers and zero, the decimal system, and the ways in which the numbers and zero are represented. The Chinese abacus gradually spread to Japan, Korea, India, the United States, and Southeast Asian countries.

The abacus is still used as a counting aid, and in addition and subtraction calculations it can be faster than electronic calculators.

Weiqi (Go)

Weiqi, or Go, is a popular board game that originated in China. It may have evolved from a method of divination practiced in China more than 3,000 years ago, in which black and white pieces were cast onto a square board marked with various symbols. *Weiqi* also involves black and white pieces on a board, but players deliberately place them on intersections of lines while trying to surround more territory than the opponent. It became popular in the Spring and Autumn Period, and prevalent in the Warring States Period.

Weiqi is played with 181 black and 180 white stones (flat, round pieces) on a square wooden board divided by 19 vertical lines and 19 horizontal lines to form 361 intersections. Each player in turn (black moving first) places a stone on the point of intersection of any two lines, after which that stone cannot be moved. Players try to conquer territory by completely enclosing areas with boundaries formed of their own stones. A stone or a group of stones belonging to one player can be captured and removed from the board if it can be completely enclosed by his opponent's stones. Groups of stones are in effect invulnerable if they contain an "eye," which consists of two or more vacant points arranged such that

Noble Woman Playing Weiqi, a Tang Dynasty painting on silk.

Women playing weiqi—an illustration from the Qing Dynasty novel *Jing Hua Yuan*

the opposing player cannot place his stone on one of the points without that stone itself being captured. Though the rules and pieces of *weiqi* are so simple that children can play, the game is considered intellectually rigorous, with billions of possible play sequences. Many experts regard it as the finest example of a pure strategy game.

Ma Rong of the Eastern Han Dynasty writes in his *Ode to Weiqi*: "The game of *weiqi* follows the rules of war. The small board is like a battlefield, where the two opponents display their troops. A slowpoke will score no gains, and a weak player will lose the play first. A coward will score no gains, and a greedy player will lose the game first." The poet sums up the strategy and tactics of the game: The player should not be too timid and too conservative, nor should he be imprudent and make rash advances. Quite a few Chinese monarchs, statesmen, and generals loved playing *weiqi*, and drew wisdom from the game, notably Cao Cao of the Three Kingdoms Period, and Emperor Xuanzong of the Tang Dynasty.

The board, pieces, and rules of *weiqi* have remained the same since ancient times, but there have been numerous manuals and works about the strategy of the game. *Weiqi* spread to India and Nepal probably in the Eastern Han Dynasty, and later to Japan and Korea, and then to Europe. The game has now reached other parts of the world.

Hot-Air Balloon

Legend has it that in the second century BC someone made an experimental hot-air balloon out of an empty eggshell. He emptied the egg white and yolk from a hole in the shell, and put a piece of burning wormwood inside the shell. As the air inside the shell was warmed, the eggshell was lifted by the wind and rose in the air. The principle of the story was quite correct

Kongming's Light (hot air balloon), invented by Zhuge Liang for use in battles.

and imaginative, and some people at that time seemed to have noticed the phenomenon of the buoyancy of hot air, which is lighter than cool air. However, the experiment could not have been a success, for even if all the space in the shell were hot air, the buoyancy would not have been sufficient to lift the eggshell. According to historical records, the inventor of the hot-air balloon was Zhuge Liang (181–234), a noted politician and strategist of the Three Kingdoms Period.

When commanding troops at the front, Zhuge Liang's health broke down from constant overwork. Before his death, he designed a light to puzzle the enemy: an oil lamp was installed under a large paper bag, and the bag floated in the air due to the lamp heating the air. These strange airborne lights are said to have frightened the enemy, who thought some divine force was helping him. Later people named this paper hot-air balloon Kongming's Light (Zhuge Liang was also known as Zhuge Kongming). In the Five Dynasties Period, a female warrior named Xin made a huge Kongming's Light using burning pine resin for use as a military signal. Fan Chengda, a poet of the Southern Song Dynasty, writes about Kongming's Light in a poem: "The candle shoots into the air and stays there." In the Yuan Dynasty, the hot-air balloon became popular throughout the country, and such balloons were launched during festivals, attracting huge crowds. Joseph Needham noted that the invention of paper in China was several centuries earlier than its use in other countries—with paper people made lanterns, and some lanterns with a very small hole in the upper part would rise and even float in the air due to the strong light and heat.

Parachute

The earliest primitive parachute was made some 4,000 years ago, when the Chinese noticed that air resistance would slow a person's fall from a height. According to the *Historical Records* by Sima Qian of the Western Han Dynasty, Shun, a legendary monarch in ancient China, was hated deeply by his father, a blind old man. When Shun was working on top of a high granary, his father set fire to the granary from below, intending to kill Shun. Holding

Shun jumping from a high granary and landing safely using two bamboo hats.

Modern parachute.

two cone-shape bamboo hats in his hands, Shun is said to have flown down and landed safely. This must be one of the earliest records of a human attempt to use a parachute-like device. Yue Ke (1183–c. 1242), a grandson of the patriotic general Yue Fei (1103–1142) of the Southern Song Dynasty (1127–1279), also recorded a similar episode in one of his essays. He writes: "There was a very high mosque in Guangzhou. One day, people found that one leg of the golden rooster at the top of the mosque had been stolen, and they wondered how the thief could have managed to escape with such a big piece of gold. When the thief was caught, he confessed that he had used two umbrellas to descend from the top of the mosque: the umbrellas had served as a kind of parachute."

In the seventeenth century, a French ambassador to Thailand also wrote about how a Chinese acrobat used umbrellas as a parachute: "The acrobat, with two umbrellas tied to his waist, jumped through an iron-wire loop high above, and the wind would bring him to the ground, on a tree, or down in the river."

In Europe it was Leonardo da Vinci (1452–1519) who first proposed the use of parachute in 1483. The first person to demonstrate the use of a parachute in action was Louis-Sébastien Lenormand of France in 1783, by jumping from a tree with two parasols.

Archery

Archaeological discoveries have shown that archery in China dates back 20,000 years. Practical archery requires three elements: a bow strong enough to propel arrows; arrows that are sharp enough to kill; and a technique to ensure the stability of arrows in flight. The bow and arrow of ancient China fully met these three conditions. Archaeologists have unearthed finely made arrowheads in a Paleolithic site in Shanxi Province. Made of stone, the arrowheads were sharp and pointed, and could be mounted on a shaft. No bow was found at the site, but since bows were usually made of wood, bamboo, and perhaps tendons

Use of the feet in archery.

of animals, these may have rotted away. The arrowheads alone were enough to indicate the existence of bows.

As for how to keep the arrow stable in flight, *Kao Gong Ji*, the earliest work on science and technology in China, writes: "Decide the proportions of the shaft to install the feathers. The feathers at the end of the shaft are installed in three directions, and then the arrowhead is mounted. An arrow thus made will not lose its balance even in strong winds." It also states: "When the feathers are too many, the arrow will slow down; when the feathers are too few, the arrow will become unstable." Later on, the ancient Chinese developed bronze arrowheads and the crossbow, upgrading archery to a new level.

Match

The match was invented in China more than 1,500 years ago. In 577, during the Northern and Southern Dynasties, the Northern Zhou (557–581) and Chen (557–589) joined forces to attack the city of Ye, capital of the Northern Qi (550–577). A long siege rendered the Imperial Palace of Northern Qi in dire need of supplies, including the facility to make a fire. Then the women in the palace invented a kindler, named *fa zhu*. *Fa zhu* was a piece of wood dipped in melted sulfur, and then dried. Rubbed against a hard board several times, the *fa zhu* would burst into flame. Tao Gu, a writer of the Song Dynasty, wrote about another igniter known as *huo cun* in one of his essays. *Huo cun* was also a piece of wood soaked with melted sulfur, but it was not easy to ignite by friction. Later on, people applied phosphor to *huo cun*, making it easy to ignite.

Outside China, an igniter similar to *fa zhu* and *huo cun* was made in the sixteenth century. The modern match was invented by the British, and later improved by the Swedish to become the safety match. This must be scratched against a

Maids of honour in the Northern Qi Dynasty are credited with the invention of the match.

surface covered with phosphor before it will ignite. The match is also called *yang huo* (foreign fire) in some Chinese dialects, but in fact the first match was invented in China more than a thousand years ago.

Chinese Kung Fu

Chinese *kung fu* (Gongfu), or Chinese martial art, is a sport peculiar to China; it combines physical exercise with personal combat skills. The martial art originated from labor, and developed later out of the needs of fighting and training. Over many years of development, Chinese martial art has generated various styles and systems. The most famous among these are: Shaolin Boxing, Taiji Boxing, Xingyi Boxing, and Drunken Boxing.

Shaolin Boxing originated and prospered at Shaolin Temple, a Buddhist temple in the Songshan Mountain, Henan Province, central China. The temple was first built in 495, and was reduced to ruins several times. But Shaolin Boxing prospered throughout. With simple, firm, and vigorous movements, this form of martial art was adopted by Buddhist monks and their followers, as well as by ordinary people. It includes several hundred formulas of fighting, such as using the fist, saber, staff, and sword. Shaolin Boxing is quite popular throughout China.

Taiji Boxing, originally known as Chen's Boxing, was invented in the fifteenth century by Chen Wangting in Chenjiagou Village, Wenxian County, Henan Province. Chen developed this form of martial art in a village context, integrating it with the Boxing Classic of Qi Jiguang (1528–1587), a general of the Ming Dynasty known for his heroic deeds in fighting Japanese bandits who invaded China's coastal areas. Later masters of Taiji adopted the theory of yin and yang and the five elements from the *Book of Changes*, making it the most popular form of martial art, characterized by deliberately slow, rhythmic movements that are circular and continuous. Chinese statesman Deng Xiaoping (1904–1997) once wrote an inscription: "Taiji Boxing is good," in order to promote this sport or exercise.

Man practicing Taiji Boxing.

Xingyi Boxing was developed in Shanxi Province of northern China in the late Ming to early Qing Dynasties. Based on the theory of mutual promotion and restraint of the five

Bruce Lee, Chinese *kung fu* star.

elements, Xingyi Boxing referred to the images and movements of twelve animals: dragon, tiger, monkey, horse, alligator, rooster, sparrow-hawk, swallow, snake, pigeon, deer, and bear. Its formulas and movements are simple, well organized, and fit for both attack and defense; they are highly rhythmic, with outbreaks of force. In martial art circles, this style is highly regarded.

Drunken Boxing, said to have begun in the Spring and Autumn and the Warring States Periods, is a form of imitation boxing. It imitates the movements of a drunken person, with the head waving constantly, the fist as swift as a shooting star, the waist as flexible as wickerwork, and the steps quick and unsteady. While staggering, the boxer moves here and there, avoiding incoming strikes, and launching his own attacks. Using quick eyesight and swift movements of hands and feet, sometimes imitating the movements of animals, the effect can be most impressive.

Football

China is the birthplace of football; the most popular game in the world, though the Chinese football team has often disappointed football fans in international competitions. In ancient China, football was called *cu ju* (kick ball). Inscriptions on Shang Dynasty bones and tortoise shells include the characters for *cu ju*, and descriptions of such a game. The sport was also quite popular in the Warring States Period. In the *Biography of Su Qin* in the *Historical Records*, historian Sima Qian wrote that residents of Linzi, the capital of Qi State, loved playing musical instruments and chess games,

The Emperor Taizu Playing Football, a copy of a Song Dynasty original, by Qian Xuan of the Yuan Dynasty.

and kicking balls. Ge Hong (c. 281–341), a noted doctor of the Jin Dynasty, in his *Miscellanea of the West Capital* wrote a story about football: After Liu Bang (c. 256 or 247–195 BC) became the first emperor of the Han Dynasty, he brought his father to the capital and treated him with all kinds of luxuries. But his father was not happy, for he missed the life in his hometown, where he could buy wine, enjoy cockfighting, and play football. To make his father happy, Liu Bang had a new Fengcheng town built after his hometown Fengyi, and moved all his friends and relatives from Fengyi to Fengcheng, so that his father could play football with his old playmates. The first book on football was also published in the Han Dynasty. At that time the ball was made of leather and hair: several pieces of leather were sewn together into a cover, and filled with a mass of hair. The first inflated football was made in the late Tang Dynasty, with an animal's bladder inside the leather cover. In the Song Dynasty, the cover of the football increased from 8 to 12 pieces of leather.

With the improvements in ball design, the field and goal for the game also changed. In the Tang Dynasty the field became formal, and the goals made taller. In the Song Dynasty, a single goal was used instead of six goals, and the participants divided into two opposing teams, each with 12 to 16 players. Playing on either side of the goal, they passed the ball between them and finally to the head player (similar to the center forward), and the ball was then shot through the goal. The ball was not allowed to touch the ground. Scoring a goal would win a chip, and three or five chips would end the game.

Numerous historical records suggest that *cu ju* in ancient China is the origin of modern football, and that China is the birthplace of the game. Joseph S. Blatter, now the president of FIFA, noted in April 1980 when he served as FIFA's technical director, that football originated in China, and spread to the West.

Tang Dynasty women playing the ball game *bu da*.

Golf

Chui wan (strike pellet), one of a number of colorful ball games of ancient China, is believed to be a precursor of the modern outdoor game of golf. *Chui wan* was originally called *bu da* (walk and hit), a game in which the player scored points by hitting the pellet into a hole in the ground. The game was developed from *cu ju*, the early form of football. Wang Jian, a poet of the Tang Dynasty, describes the game in one of his poems: "Stands have been set up on both sides of the palace hall, during the Hanshi festival court people play the *bu da* ball. They walk and kneel in competition, and the champion thanks the Emperor when he wins." This suggests that *bu da*, a game similar to golf, was played in China more than 1,000 years ago.

In a painting from the Song Dynasty two children are shown playing *chui wan*, and in the Yuan Dynasty, *Wan Jing* (Classic of Chui Wan) was published. This book says Emperor Huizong of the Song Dynasty and Emperor Zhangzong of the Jin Dynasty were both enthusiastic players of *chui wan*. A Ming painting, *Ming Xuanzong Xingle Tu* (*Emperor Xuanzong of the Ming Dynasty*

on a Pleasure Ride), shows the Emperor, in plain clothes, striking the pellet in the field. The painting also shows the course, the cup, and colorful banners. The club, ball, course, and rules of *chui wan* were similar to those of modern golf. In Europe, the game of golf was first seen in paintings of the fourteenth and fifteenth centuries—several hundred years after the Chinese game of *chui wan*.

Classical Works of Science

Scientific literature and archeological discoveries play an important role in the study of the history of science, and ancient classics of science and technology have helped researchers rediscover a large number of lost inventions and discoveries. The following are the ten most important works from among the ancient Chinese scientific classics now available to readers.

Kao Gong Ji
(Records of Examination of Craftsmen)

This is an important work of science and technology and one of the classics before the Qin Dynasty. Its author is unknown. Scholars have established that this book was an official document of the Qi State of the Spring and Autumn Period. It was used to guide the handicraft industries, including the examination and assessment of craftsmen, including carpenters, metalworkers, leatherworkers, porcelain makers, and many others. It also records the manufacture of vehicles, weapons and musical instruments, and the building of houses.

Pages from *Kao Gong Ji* (*Records of Examination of Craftsmen*).

Huangdi Nei Jing
(The Yellow Emperor's Canon of Internal Medicine)

This, the earliest extant medical classic, consists of two parts: *Su Wen* (*Plain Questions*), and *Ling Shu* (*Miraculous Pivot*). It was

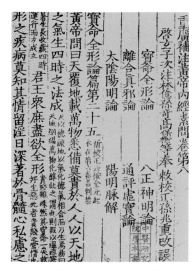

Page from *Huangdi Nei Jing* (*The Yellow Emperor's Canon of Internal Medicine*).

probably compiled from before the Qin Dynasty to the Western Han Dynasty, and was written under the name of the Yellow Emperor. The work summed up the experience and theoretical knowledge of ancient people in fighting diseases, and laid some of the foundations for the development of Chinese medicine. *Su Wen* mainly deals with basic theories of medicine, while *Ling Shu* discusses acupuncture. Based on primitive dialectical materialism, the work elaborates human physiology and pathology, yin and yang, the five elements, the correlations between man and nature, the internal organs and viscera, the channels and collaterals, diagnostic methods, differentiation of symptoms and signs, the principles of administering drugs, and acupuncture. It also develops for the first time the idea of the circulation of the blood. The work lists the symptoms and signs of 310 diseases, covering internal medicine, surgery, and gynecology. As an excellent example of the integration of basic theory and medical practice, the work is still used by contemporary practitioners of traditional Chinese medicine.

Shanghan Zabing Lun
(Treatise on Febrile and Miscellaneous Diseases)

This medical classic was written by Zhang Zhongjing (c. 150–219) in the later years of the Eastern Han Dynasty. It analyses and differentiates febrile diseases according to the theory of six channels, and miscellaneous diseases according to the pathological changes of viscera and bowels and their interrelations. In so doing it establishes Chinese medicine's

Zhang Zhongjing.

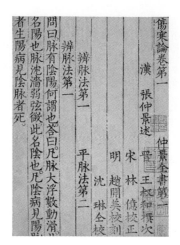

Page from *Shanghan Zabing Lun*
(*Treatise on Febrile and Miscellaneous Diseases*).

theoretical system and a therapeutic principle—diagnosis and treatment based on an overall analysis of signs and symptoms, and thus helps to lays the foundations for the development of clinical medicine. Later generations divided it into two books. One is the *Treatise on Febrile Diseases*, in which there are more than 40 diseases and 200 prescriptions; the other is *Essential Prescriptions of the Golden Coffer*. They contain, basically, the prescriptions often used in every department of clinical medicine and are known as the earliest reference works dealing with prescriptions. The book provides some scientific methods for emergency medicine, such as treatment of attempted suicides, and artificial respiration. For the preparation of drugs, the book lists decoction slices, powder, alcoholic solution, lotion, fumigant, ear and nose drops, suppository, enema, and ointment.

Qi Min Yao Shu
(Main Techniques for the Welfare of the People)

This earliest intact work on agriculture in China was written by Jia Sixie of the Northern Wei Dynasty (386–557). Jia was an

田則畝收十石其美與蓋矢熟糞同凡秋收之

月中穫泫泆反種七月八月犁掩穀之爲春穀

美田之法綠豆爲上小豆胡麻次之悉皆五六

宜縱牛羊踐之根蹉復秋耕之則死菅茅之地初

耕欲深轉地欲廉勞欲再春再

秋耕掩青者爲上比至冬月青草復生乃至其

夏欲淺犁掩一歲青者之間凡秋耕欲深春夏耕

之天時地利若不牛勞而不草木不蕃秋田暴言澤遇

Page from *Qi Min Yao Shu* (*Main Techniques for the Welfare of the People*).

official, and also a manager of agricultural production. Written in about 533–544, the book consists of ten volumes, with 110,000 Chinese characters. Its ninety-two chapters discuss the cultivation of grain crops, vegetables, fruits, bamboo, and trees, the raising of livestock, poultry and fish, the processing of farm products, the making of wine and liquor, and other subsidiary operations. The book summarizes the practical aspects of agricultural production in the middle and lower reaches of the Yellow River. It demonstrates that agricultural production in China was quite advanced at that time: dry farming, grafting of pear trees, propagation of saplings, castration of livestock and poultry for fattening, and farm produce processing techniques.

Mengxi Bitan (Dream Pool Essays)

Written by Shen Kuo (1031–1095) of the Northern Song Dynasty, these essays were written at Shen's residence of Mengxi (Dream Pool) Garden in Runzhou (now Zhenjiang, Jiangsu Province)—hence the name *Dream Pool Essays*. The 609 items covered in thirty volumes of the book include astronomy, mathematics, physics, chemistry, biology, geology, geography, meteorology, medicine, agriculture, engineering, literature, history, music, and the fine arts. In the natural sciences section, the author recorded scientific achievements in ancient China, especially in the Northern Song Dynasty, such as Bi Sheng's movable type

Shen Kuo.

printing, and advanced techniques of iron and copper smelting. He described the application of petroleum, and used for the first time the term *shi you* (literally meaning "stone oil," the Chinese for petroleum). The author also explained some natural phenomena, for example: rainbows come about due to refraction of sunlight; the tide is related to the movement of the moon; the main content of meteorites in Changzhou is iron; and the earth's surface is formed by erosion of water. Shen was the first to discover magnetic variation, and he also discussed refraction of rays by a lens, and resonance.

Yingzao Fashi (Architectural Standards)

This is a monograph on architecture published in the Northern Song Dynasty. Compilation of the book was started during 1068–1078 by the government department in charge of construction,

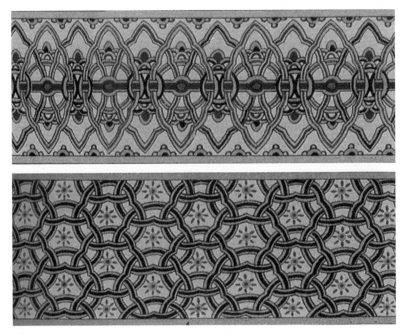

Colourful patterns from *Yingzao Fashi* (*Architectural Standards*).

The Anping Bridge in Quanzhou, Fujian Province. Built between 1138 and 1151, this stone bridge is the longest surviving ancient seaport bridge in China.

and finished in 1091. Later it was edited by Li Jie (?–1110), finished in 1100, and republished in 1103. The work comprises 34 chapters with 357 articles and 3,555 items, covering the terms of buildings, management of various workers, time limits for different tasks, materials, and patterns and designs. It discusses in detail the jobs of masonry and carpentry, as well as those involving bamboo and soil. It also includes detailed processes and specifications for the construction and painting of buildings, including the materials and labor required, as well as indicating flexibility in design to ensure the quality of buildings. Much of its content came from the experience of skilled workers and craftsmen. This work provided standards for official buildings in the central part of China, and as it contains a number of patterns, designs and data on ancient buildings and decoration, it is also of great significance for the study of the history of Chinese architecture.

Nong Shu (Treatise on Agriculture)

This comprehensive work on agriculture was written in the Yuan Dynasty by Wang Zhen, who twice served as county magistrate. It is a guide to local agricultural production, based on lengthy studies, and was finished 1313. The book consists of three parts: the first part deals with various aspects of agriculture in general; the second part looks at the cultivation of various crops, fruits, vegetables, bamboo, and trees; while the third and major part catalogues farm tools and other instruments, with more than 270 illustrations. As some of the tools have long been lost, the

Yang ma (seeding horse), a device to take the strain out of transplanting rice seedlings. Illustration from Wang Zhen's *Nong Shu* (*Treatise on Agriculture*).

records in this book are invaluable. In many places the author makes a comparison between farming in north and south China, including differences in farm tools used, and their advantages and disadvantages. In an appendix, Wang Zhen tells readers about the movable type printing technique used to print this work, based on earlier inventions, but improved by the author himself.

Bencao Gangmu (Compendium of Materia Medica)

Written by Li Shizhen (1518–1593) in the Ming Dynasty and completed in 1578, this work, in 52 volumes, is divided into 16 units and 60 classes, with 1.92 million characters. It contains 1,892 kinds of drug, including some 300 to 400 discovered by the author himself, and classified in a systematic way. For each drug the author first states its name and place of origin, then provides an illustration, and discusses methods of cultivation

Li Shizhen.

or collection. He then verifies real varieties from fake ones, and corrects some mistakes in previous medicinal classics. He also discusses the processing, properties, and functions of the drugs. The author collected over 11,000 formulae and prescriptions, including those used by ancient physicians and other effective remedies used by ordinary people from generation to generation. He also provides 1,100 illustrations of the drugs. This work summarized the achievements in Chinese *materia medica* and botany before the sixteenth century. It was first published in 1596, and translated into Japanese in 1607 and there are now also English, French, German, and Russian versions of this great work. It was also known to British scientist Charles Darwin (1809–1882), who praised it highly.

Tiangong Kaiwu
(The Exploitation of the Works of Nature)

Authored by Song Yingxing (1587–1661) of the Ming Dynasty, this work was first published in 1637, in eighteen volumes in three parts. It systematically records agricultural production and techniques and the handicraft industry in ancient China, with 121 illustrations. The first volume covers cultivation of grain, cotton and other fiber crops, rearing of silkworms, spinning, dyeing, food processing, and the manufacture of salt and sugar. The second volume deals with the production of bricks and tiles, porcelain, iron and steel, ships and vehicles, limestone and lime, coal, sulfur, edible oil, candles, and paper. The third volume discusses mining of minerals, smelting of metals, production of weapons, gunpowder, ink, dyes, yeasts, and the manufacture of pearls and jade articles. The three volumes

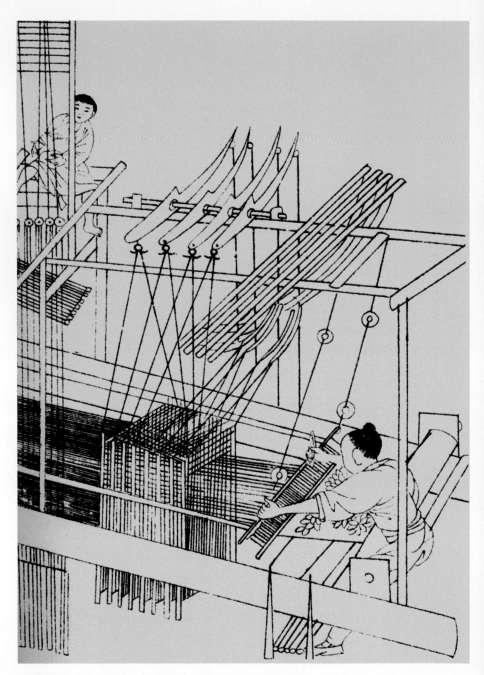

Illustration of a loom, from *Tiangong Kaiwu* (*The Exploitation of the Works of Nature*).

outline 130 production techniques, detailing types of materials, the quantities to be used, their origins, the structures of the tools needed, and the production processes. The author describes most of the techniques in use at that time. Soon after its publication, the work caught the attention of people in other countries and was translated into several languages, but was later lost in China until 1920, when a Chinese edition was published from a back translation of the Japanese version.

Nong Zheng Quan Shu (Complete Treatise on Agriculture)

This work was written by Ming Dynasty official Xu Guangqi (1562–1633), who was born in Songjiang, Jiangsu Province, an area where agriculture was quite well developed. Xu had a deep love of agriculture, and during his three-year leave of mourning his father, he carried out extensive local experiments in farming, and later went to Tianjin several times for further research, which helped him complete this encyclopedic work. Xu's great work

was published posthumously in 1639, with the help of his friends. The work was divided into 60 volumes, with more than 700,000 characters. It covers farming, farmland management, water conservation, farm tools, forestry, mulberry cultivation and raising silkworms, animal husbandry, manufacturing, and famine relief. In addition to content based on his experience in cultivation of grain and cotton crops, Xu

The Italian Jesuit missionary Matteo Ricci and Xu Guangqi.

devoted much space to water

Farm workers.

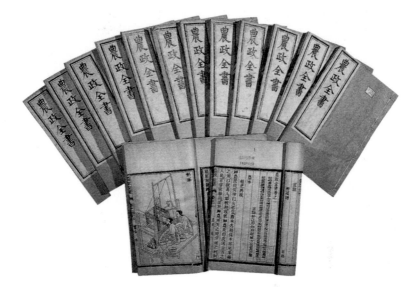

Nong Zheng Quanshu (Complete Treatise on Agriculture).

conservation and famine relief. He studied the famine years in Chinese history, including 111 locust plagues. He held the view that to fight famine it was necessary to build water control facilities. Thus, to fight famine in the northwest, he stated that it would be better to build water conservation facilities and reclaim wasteland there to expand grain production than to transport grain supplies there at high cost. In addition to the author's own views, this great work of agricultural science also drew heavily on the literature of the Ming and previous dynasties.

Appendix:
Chronological Table of the Chinese Dynasties

The Paleolithic Period	c. 1,700,000–10,000 years ago
The Neolithic Period	c. 10,000–4,000 years ago
Xia Dynasty	2070–1600 BC
Shang Dynasty	1600–1046 BC
Western Zhou Dynasty	1046–771 BC
Spring and Autumn Period	770–476 BC
Warring States Period	475–221 BC
Qin Dynasty	221–206 BC
Western Han Dynasty	206 BC–AD 25
Eastern Han Dynasty	25–220
Three Kingdoms	220–280
Western Jin Dynasty	265–317
Eastern Jin Dynasty	317–420
Northern and Southern Dynasties	420–589
Sui Dynasty	581–618
Tang Dynasty	618–907
Five Dynasties	907–960
Northern Song Dynasty	960–1127
Southern Song Dynasty	1127–1276
Yuan Dynasty	1276–1368
Ming Dynasty	1368–1644
Qing Dynasty	1644–1911
Republic of China	1912–1949
People's Republic of China	Founded in 1949